Solar Fuels
Materials, Physics, and Applications

T0139135

Solar Fuels
Materials, Physics, and Applications

Theodore Goodson III

CRC Press
Taylor & Francis Group
Boca Raton London New York

CRC Press is an imprint of the
Taylor & Francis Group, an **informa** business

CRC Press
Taylor & Francis Group
6000 Broken Sound Parkway NW, Suite 300
Boca Raton, FL 33487-2742

First issued in paperback 2022

ISBN-13: 978-1-439-87491-2 (hbk)
ISBN-13: 978-1-03-233966-5 (pbk)
DOI: 10.1201/9781315374635

Library of Congress Cataloging-in-Publication Data

Names: Goodson, Theodore G.
Title: Solar fuels : materials, physics, and applications / Theodore Goodson III.
Description: Boca Raton : CRC Press, [2017] | Includes bibliographical references and index.
Identifiers: LCCN 2016051830| ISBN 9781439874912 (hardback : alk. paper) | ISBN 9781439874936 (ebook)
Subjects: LCSH: Solar batteries. | Solar energy. | Renewable energy sources.
Classification: LCC TK2960 .G66 2017 | DDC 621.31/244--dc23
LC record available at https://lccn.loc.gov/2016051830

**Visit the Taylor & Francis Web site at
http://www.taylorandfrancis.com**

**and the CRC Press Web site at
http://www.crcpress.com**

To my wife Stephanie Lynn
and my kids Jared (Theodore IV), Elizabeth, and Sean

Contents

List of Figures

Preface

In this book, I have attempted to provide an interesting view about solar energy production for both new and experienced scientists and engineers, as well as for enthusiasts who have a little technical background. The book is thus not intended to provide complete analytical discussions as a textbook would. Instead, I provide the key concepts and results and subsequent approaches in these areas. The book is designed for study in general for all interested thinkers to enjoy the subject without extensive background in mathematics and physics. However, the details are provided in a concise manner and the references therein will guide the reader in the direction toward learning more. Great effort has been made to make the book both very recent in content and practical in approaches and subject matter. Even more to this end, I have attempted to capture the most recent thoughts and predictions about the effect of this field on our nation's (and the world's) energy economy and market place.

In Chapter 1, the book provides a substantial introduction to the principal issues behind the goal of increasing the use and efficiency of solar cell devices in our energy economy. This chapter also outlines the basic types of solar cell devices and their positive and negative characteristics presently limiting their use in real translatable devices. This sets up chapters related to the mechanism of how present organic solar cells work, the critical organic structures used in the devices, how we measure the important parameters with a large variety of experimental techniques, modernization of organic cell design, the importance of the interfaces in organic solar cell devices, and new approaches to beat old limits to solar cells. Chapter 6 presents newer approaches with singlet exciton fission as well as with organometallic perovskite materials. In Chapter 7, I summarize and provide an outlook into what might be on the horizon for this field. The goal of these chapters is to ultimately be used as a reference point for the reader to learn the basics of the topic and to be able to come back to this topic again and remember what has already been accomplished and what are the present limitations in the field.

After much thought, I decided to concentrate mainly on organic solar cell devices. The field of solar development with inorganic devices as well as with silicon enjoyed a great deal of success and attention in other well-written texts in the past. The introductory chapters briefly review some of the approaches and successes with inorganic solar materials. The extensive reference list is straightforward and will help in finding out more concerning the physics and chemistry of these organic devices.

I am very much indebted to colleagues from around the world for the detailed discussions in this field for the preparation of writing this book. I thank Professor Luping Yu for his continued encouragement and expert advice on this subject, Professor Victor Batista for his close attention to detail and suggestions, and Professors Peter Green and Mike Wasielewski for their expert advice and work. Finally I would like to thank Pamela and Bruce Epstein as well as my parents (Exie and Theodore Goodson) for their encouragement in finishing and proofreading this book.

Author

Theodore Goodson III received his BA in 1991 from Wabash College and earned his PhD in chemistry at the University of Nebraska-Lincoln in 1996. After postdoctoral positions at the University of Chicago and at the University of Oxford, he accepted a position as assistant professor of chemistry at Wayne State University in 1998. In 2004, he moved to the University of Michigan as professor of chemistry. In 2008, he was appointed as the Richard Barry Bernstein Professor of Chemistry at the University of Michigan. Dr. Goodson's research centers on the investigation of nonlinear optical and energy transfer in organic multichromophore systems for particular optical and electronic applications. His research has been translated into technology in the areas of two-photon organic materials for eye and sensor protection, large dielectric and energy storage effects in organic macromolecular materials, and the detection of energetic (explosive) devices by nonlinear optical methods. He has investigated new quantum optical effects in organic systems that have applications in discrete communication systems and sensing. Dr. Goodson's lab was also the first to investigate the fundamental excitations in small metal topologies that are now candidates for tissue and other biological imaging. In 2009, he founded Wolverine Energy Solutions and Technology Inc., a start-up company with contracts to produce high-energy-density capacitors for military, automotive, and medical devices. The company also developed a new system for the detection of IEDs remotely, with one of the patents awarded to Dr. Goodson at the University of Michigan. His awards include the Distinguished University Faculty Award, the National Science Foundation American Innovation Fellowship, the Research Young Investigator Award, National Science Foundation CAREER Award, Alfred P. Sloan Research Fellowship, the Camille Dreyfus Teacher-Scholar Award, the Lloyd Ferguson Young Scientist Award, the Burroughs Welcome Fund Award, the American Chemical Society Minority Mentorship Award, the University Faculty Recognition Award, the College of Science Teaching Award, and a National Academy of Sciences Ford Postdoctoral Fellowship. Dr. Goodson has been a senior editor for *The Journal of Physical Chemistry* since 2007. Dr. Goodson has published more than 150 scientific publications.

1 Historical Background and Structure of This Book

IN THE BEGINNING

Solar energy is the oldest form of natural energy on planet Earth (Figure 1.1). Its seemingly unlimited power has mystified cultures who have praised Ra and the Phoenix.[1] Once thought too powerful or too dangerous to harness, the sun has moved into the spotlight finally to offer a real solution to our energy needs. The use of the sun's heating potential can be traced back as far as the first century AD[2] where sun rooms appeared in Roman architecture. And as early as the 1700s, complete designs and construction of the first solar collectors were created in order to do actual work as well as cook food.[3] Indeed, the idea of utilizing the sun's energy for all its worth is not new. After all, it was in 1816 that this first solar thermal electric technology appeared, which concentrates the sun's thermal energy in order to produce power.[4] Science, ingenuity, inspiration, and timing appear to be the critical elements toward the pioneering developments in the use of solar energy. As in many paradigm shifts in technology, there comes a time when the summation of these elements comes to an uplifting cadence and the world realizes the need for something new and something better that will maintain our way of life and provide for those in the future. It has happened before; for example, the inspiration of famine and wars has galvanized the creation of new technology that has both saved and enlightened our way of living. Today's inspiration is a result of diminishing fossil fuels and the economics of a threatening new world.[5]

While the idea of utilizing solar fuels is not new, the development of the materials and the specific mechanisms that might ultimately provide the best solution have come a long way since the early days of solar thermal electric technology. Indeed, the creation and perfection of solar panels have enjoyed great success. A solar cell is any device that directly converts the energy in light into electrical energy through the process of photovoltaics.[6] The development of solar cell technology begins with the 1839 research of French physicist Antoine-César Becquerel.[7] Becquerel observed the photovoltaic effect while experimenting with a solid electrode in an electrolyte solution when he saw a voltage developed when light fell upon the electrode.[7] This photovoltaic mechanism provided the much needed insight into a possible strategy of producing useful and possibly efficient energy from the sun. It is believed that in 1883, the first functioning solar cell was made by Charles Fritts,[8] who used junctions formed by coating selenium (a semiconductor) with an extremely thin layer of gold. However, early solar cells had energy conversion efficiencies of less than 1%.[9]

FIGURE 1.1   **(See color insert.)** The gods of ancient Egypt—Aten and Ra. Ra in the solar bark. (Credit: www.shutterstock.com, 122013538.)

In 1941, the silicon solar cell was invented by Russell Ohl.[10] And in 1954, three American researchers, Gerald Pearson, Calvin Fuller, and Daryl Chapin, designed a silicon solar cell capable of a 6% energy conversion efficiency with direct sunlight.[11] This, for the first time, gave those concerned with our world's energy economy great attention as this suggested that with further development, silicon solar cells were indeed a viable energy provider. In this invention, an array of several strips of silicon (each about the size of a razorblade) placed in sunlight captured the free electrons and turned them into electrical current.[11] They had created the first solar panels. As will be seen in this analysis of the critical points of organic solar cell technology, new materials and devices are now closer to 10% efficiency with good reproducibility.[12] It still might be a hard sale to replace silicon. The present silicon solar cells operate in the range of 15%–25% depending on the scale and operating conditions.[13] They absorb nearly 60% of solar light and their small bandgap (1.1 eV) enables photons with low energy (e.g., red light) to create the necessary electron–hole pairs that generate photocurrent.[13] However, the absorption of high-energy photons (blue light) results in "hot" electrons, which are both electronically excited and thermally activated[14] and lose most of their energy as heat without contributing to electric power.[14] Therefore, a hurdle to boosting energy conversion efficiency in organic photovoltaic solar cells is not just replacing those made from silicon, but it is also to capture the excess energy of the thermally unrelaxed electrons before it is lost as heat.

GOALS FOR FUTURE SOLAR CELLS

In discussing the relatively broad topic of solar fuels, it is very important to know the goals of such a massive undertaking. As discussed briefly earlier, there is a long

FIGURE 1.2   (See color insert.) Power plant using renewable solar energy from the sun. (Credit: www.shutterstock.com, 177900254.)

history of the use of solar energy for various important applications (Figure 1.2). So it is necessary to specify particular goals for a modern analysis of this area. Thus, a major goal in this undertaking is related to the development of new materials for the construction of modern solar cells. The materials used in modern solar cells may be divided into two parts. Historically, it has been inorganic materials that first arrived on the scene in the construction of modern photovoltaic devices.[15] The use of molecular beam technology has allowed a great degree of success in this methodology as it has allowed the precise deposition of the inorganic materials on substrates to very high resolution.[16] The other kind of material used in modern solar cell development is that made of organic (made of primarily oxygen, nitrogen, carbon, and hydrogen) systems.[17] In this area of research, a virtual explosion of interest and investments has recently come to the focus of technology. From basic research to the translation of novel materials and devices, the area of organic solar cells is at the penultimate step in the development of competitive and productive materials that can one day be commercialized. Not only has the field of organic solar cell discovery enjoyed a number of great accomplishments potentially worthy of translating into the marketplace, but the field has developed a deeper understanding of the processes and science involved in their mechanisms. In many ways, the fruit of this knowledge has already begun to expose itself in other areas of science.[18] And the inspiring ideas and creative solutions illustrated by this basic research have prompted a serious consideration about the possibility of this form of solar energy consumption one day being a real alternative solution. It is because of these reasons that a major goal of this undertaking in analyzing the developments of solar fuels will focus primarily on organic solar materials. A close look at the developments in this area, some failures and many successes, as well as looking to the future for these materials will be discussed.

MOLECULAR PROCESSES IN ORGANIC SOLAR CELLS

While the understanding of new materials is critical to one's appreciation of where the field of organic solar cells is heading, it is also important to discuss the critical lessons learned about the construction and engineering of these devices. The use of selected fabrication procedures has allotted the expertise in providing relatively efficient and reproducible results utilizing organic materials for solar devices. The earliest devices with single layers of active organic solar cell materials provided the needed standards for what was necessary in dealing with organic semiconductor solar cell materials.[19] Many of these materials were organic polymers at first, then small molecules and later other organic molecular architectures.[20] Later, after much investigation, the concept of the bulk heterojunction arrived and was initially introduced by blending two polymers having both donor and acceptor properties in solution.[21] This provided further discussions and developments in the area of mobility and diffusion of excitons and charges in such devices. The fabrication of the films and their properties became a major obstacle. A number of techniques arrived which would allow the coating of particular solutions to be homogeneous and provide less defects. For example, spin cast films from binary polymer solutions could result in solid state mixtures of both polymers with good properties but could also be optimized by the choice of polymer or particular small molecule additives.[22] The area of device and film fabrication has developed even further with other techniques such as the lamination of two polymer layers.[23] Higher power conversion efficiencies have been reported with such devices which provide a relatively diffusive interface between the donor and acceptor polymer structures. This field continues to expand in its approach toward the fabrication of solar devices utilizing solar materials.

In addition to the fascinating work carried out on bulk heterojunction organic materials, there is also considerably large effort in the field of dye-sensitized electrochemical solar cells. This area has received great attention as in its initial phase of development, there were important and very well-received accomplishments made by scientists such as Graetzel and others.[24,25] The different methods have learned from each other. For example, a number of approaches have introduced organic hole conductors in place of the liquid electrolytes in electrochemical solar cells.[26] There has also been a push to the possible exchange of the electron-conducting acceptor materials in organic heterojunction devices with inorganic nanocrystals. Thus, it appears that the electrochemical and organic photovoltaic research directions are gradually merging together in order to provide the best possible solution. Again, this puts great emphasis on the basic research nature of much of the work that has been done in organic solar cells development over the past 40 years.

EXCITONS AND ORGANIC SOLAR CELLS

The goals of understanding the materials used in solar cells and their fabrication into devices are the first steps in obtaining a basic grasp of what and where this field is at in terms of its development. One must also understand the basic physics of how electrons and holes move throughout the material and produce efficient transfer. It is well known now that in order to create a working photovoltaic cell, the two photoactive

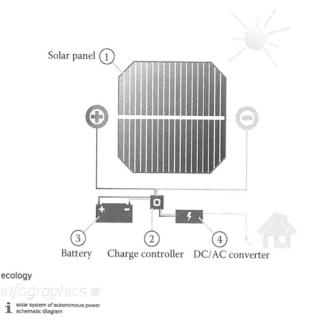

Solar panel ①

Battery Charge controller DC/AC converter
③   ②   ④

ecology
infographics 🔖
i solar system of autonomous power
   schematic diagram

FIGURE 1.3 **(See color insert.)** House equipped for the use of solar energy. (Credit: www.
shutterstock.com, 186664391.)

materials are sandwiched between two metallic electrodes to collect the photogenerated charges. Generally, one of the electrodes is metallic and the other is transparent as to allow for good solar photon capture. After the charge separation process,
the charge carriers have to be transported to these electrodes without recombination.
Finally, it is important that the charges can enter the external circuit at the electrodes
without interface problems. Thus, it is critical to understand the four basic steps in
the solar cell function (Figure 1.3). It is now generally believed that the process of
converting light into electric current in an organic photovoltaic cell is accomplished
in four consecutive steps: (1) *absorption* of a photon leading to the formation of an
excited state, the electron–hole pair (exciton); (2) *exciton diffusion* to a region; (3)
charge separation; and (4) *charge transport* to the anode (holes) and cathode (electrons).[27] The potential energy stored within one pair of separated positive and negative charges is equivalent to the difference in their respective quasi-Fermi levels, or
in other words it corresponds to the difference in the electrochemical potentials. The
larger the quasi-Fermi level splitting that remains during charge transport through
the interfaces at the contacts, the larger will be the photovoltage.[28] For ideal (ohmic)
contacts, no loss is expected, and energy level offsets or band bending at nonideal
contacts (that undergo energy-level alignments due to Fermi-level differences) can
lead to a decrease in the photovoltage.[28] The electric current that a photovoltaic
solar cell delivers corresponds to the number of created charges that are collected
at the electrodes. This number depends on the fraction of photons absorbed (η_{abs}),

the fraction of electron–hole pairs that are dissociated (η_{diss}), and finally the fraction of (separated) charges that reach the electrodes (η_{out}) determining the overall photocurrent efficiency.[29] It is the physics and chemistry of these processes that determines the overall success or failure of the device. There are also other conditions related to the device architecture that will also heavily determine its efficiency. For example, it is well known now that a higher fill factor (strongly curved J–V characteristic) is advantageous and indicates that fairly strong photocurrents can be extracted close to the open-circuit voltage.[30] In this range, the internal field in the device that assists in charge separation and transport is fairly small. Consequently, a high fill factor can be obtained when the charge mobility of both charges is high. Presently, the fill factor is limited to about 60% in the best devices, but values up to 70% have been achieved recently.[31] A firm grasp on these concepts in relation to organic solar materials is critical. Indeed, at the efficiencies at this point in time, there is still room for improvement. But it is believed that a deeper understanding of the structure–function relationships of these four basic parameters of organic solar cells may one day lead to a superior device able to compete with inorganic and silicon-based devices. A major goal of this book is for the reader to grasp the fundamental processes and to appreciate the complexity of the mechanisms involved in organic solar cells.

In connection with a good understanding of the basic physical phenomena involved with organic solar cells, it is also important to discuss an ongoing problem with the issue of excitons and their diffusion through the material. There has been great effort, both experimentally and theoretically, regarding the effects that govern this process in organic materials.[32] While this field appreciates the contributions of both chemists and physicists as well as engineers, it has been said that the language (or nomenclature) might also present some confusion in the understanding of the problem and what solutions are to make an impact. In general, it is well understood that an exciton is a bound state of an electron and hole which are attracted to each other by the electrostatic coulombic force. It is an electrically neutral quasiparticle that exists in insulators, semiconductors, and some liquids.[33] The exciton is regarded as an elementary excitation of condensed matter that can transport energy without transporting net electric charge. An exciton forms when a photon is absorbed by a semiconductor. This excites an electron from the valence band into the conduction band. In turn, this leaves behind a localized positively charged hole. The electron in the conduction band is then attracted to this localized hole by the coulomb force. This attraction provides a stabilizing energy balance. Consequently, the exciton has slightly less energy than the unbound electron and hole. In this book, we will discuss the dynamics of excitons in relation to the efficiency of organic solar cells.

In consideration of the exciton diffusion, a great deal of effort has been carried out with regard to the lifetimes and path of the exciton in the solid state device.[34] It is well known that these excitons travel a certain distance during their lifetime. This distance is called the exciton diffusion length and is a critical component in device efficiency. To produce current, an exciton must be able to diffuse and reach the interface between the donor and acceptor components during its lifetime. If the interface is, for example, 20 nm from where the exciton is generated, but the exciton only diffuses 5 nm during its lifetime, then no current will be produced. If a system has large diffusion lengths but the migration of the exciton takes too long, then the distance between where the exciton is generated and the interface must be reduced and will result in thinner films. However, thinner films absorb fewer photons. Organic cell engineers must factor in exciton diffusion lengths, exciton lifetimes, and film thickness to design efficient cells. The film must be thick enough to absorb a reasonable amount of photons but also be thin enough so that the generated excitons can reach the interface efficiently. This whole process occurs in excitonic solar cells. Indeed, a close inspection of all of these factors and how they are balanced for the best combination is important in a general understanding of how organic solar cells will proceed in the future. This book aims to discuss some of the major factors and the approaches that scientists and engineers have explored in order to control these factors (Figure 1.4).

Not only has academia played a major role in the developments of organic solar cell devices, but there have also been a number of major research laboratories and industrial scientists who have added an enormous amount of information to this subject. For example, with regard to the issue of stability, increased power efficiency in different environments and in the area of fabrication procedures, the field has enjoyed a close interaction with academic and industrial-based research in the last 30 years.[35] In terms of environmental effects, it is clear now that organic materials may suffer greater losses from typical atmospheric effects that do not affect inorganic systems as strongly. For devices made with organic polymers, earlier studies have revealed that chemical degradation of organic solar cells mainly involves oxygen, water, and electrode material reactions within the active polymer layer. Experimentalists at industrial labs and national laboratories have advanced our understanding of how these effects are critical to the operation and stability of an organic solar cell. Particular measurements have been developed for the understanding and lifetime of these effects in organic solar cells. For the measurement of the degradation of these devices, the device parameters such as the short circuit

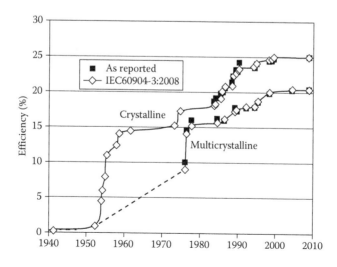

FIGURE 1.4 The evolution of crystalline and multicrystalline silicon solar cell efficiency.

current J_{SC} and the open-circuit voltage V_{OC} are determined and put into context of the structure–function relationships of the organic material. Also, a decrease in J_{SC} and V_{OC} after exposure to various environmental conditions indicates quantitatively whether the particular device degrades. However, for the determination of J_{SC} and V_{OC}, devices need to be prepared in a particular manner which was developed over a number of years. Many of these methods are directed at the characterization of the active layer as opposed to the entire device. As the reader will see in numerous chapters in this book regarding organic solar cells, for bulk heterojunction solar cells, the donor and acceptor materials are often phase separated on the nanometer scale which may present both good and bad effects in regard to the degradation of the device structure. The addition of various blocking layers has been an attractive approach toward solving this problem. This brings us to another major issue in organic solar devices, namely, characterizing the interface. Adding blocking layers may form an interface with the conjugated polymer, and it is hoped that this will have a positive effect on the degradation processes at their interface.[36] Thus, new methods allowing us to (electrically) characterize materials and interfaces at a length scale of nanometers could lead to new insights into degradation processes. As this book will demonstrate, characterizing the interface of an organic solar cell is a major obstacle that may limit the overall efficiencies of present devices.

PROBLEM AT THE INTERFACE

As it became clear in the last two decades, it is not only the chemistry of the materials that is of vital importance to the success of a solar cell, it is also the surface and film characteristics that are also important. The development of microscopy over the last three decades has seen an enormous degree of increase in resolution and methodology to perfect the imaging and scanning ability of organic materials.[37] These developments have been successfully applied to organic solar cells. One such method that has received a great deal of attention is that of scanning probe microscopy (SPM).[38] Indeed, from studies of the materials to the actual devices, it has been shown that this technique can be used as a productive tool to investigate surface topography at the nanoscale level. In addition to topography, electric properties of thin films and devices can be investigated on a nanometer scale using a variety of different modes. In the case of organic polymer–metal nanoparticle topologies, the close vicinity of a metal alters the excited deactivation pathways of the surface-bound molecules. For example, it has been observed that a distance-dependent quenching of excited states of chromophores on metal surfaces in this geometry may be obtained. One of the noticeable properties of the fluorophore molecules when bound to metal surfaces is the decrease in singlet lifetime as a result of energy transfer from excited dye molecules to bulk metal films. The total quenching of the singlet-excited state of the chromophores can limit the application of chromophore-labeled metal nanoparticles in optoelectronic devices such as organic solar cells. A better understanding of the excited-state processes will enable the effective utilization of chromophore-functionalized metal nanostructures for light harvesting and other specialized applications. Possible deactivation pathways of the photoexcited fluorophore bound to a gold

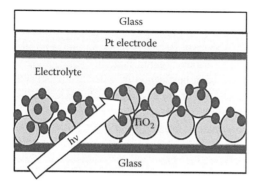

FIGURE 1.5 (**See color insert.**) Basic structure of dye-sensitized solar cell.

nanoparticle by surface and molecular sensitive techniques to probe the intermolecular interactions, energy transfer, electron transfer, and emission from the chromophores in different topologies to metal particles are needed (Figure 1.5).

MEASUREMENTS OF SOLAR CELL SURFACE PROPERTIES

A number of different forms of SPM methods have already been developed to directly approach the issues surrounding surface topography in organic solar cells. For example, the reader will learn of recent developments of time-resolved electrostatic force microscopy to study photo-oxidation and trap formation in bulk heterojunction photovoltaic materials.[39] Also, the use of a Kelvin probe force microscopy (KPFM), it was demonstrated that one can measure the local variations of work function differences between an SPM tip and a surface.[40] This powerful new technique may actually help scientists and engineers to probe the local surface potential of active layers with and without the application of light. Such sensitive measurements allow scientists in this area to simulate working conditions and allow a visual of the photo-induced charge carrier generation in an organic (possibly amorphous or blend) material. Detailed reports have appeared from the use of such sensitive tools where surface potential changes upon illumination of polymeric bulk heterojunction thin films using KPFM showed that differences in the surface potential between a nonilluminated and an illuminated sample were related to the device performance. Later, it was discovered that local photocurrents might also be able to probe the device lifetime and performance.[41] For example, the short circuit current conditions can be measured by SPM using the electrical conductive scanning force microscopy (cSFM) mode.[42] The cSFM method has been used to map local photocurrents in the donor/acceptor blend solar cells of the conjugated polymers with PCBM with great success. This was particularly important as it was found that the average photocurrent measured with cSFM agreed well with the photocurrents for bulk devices with Al contacts and can therefore be correlated with the device performances as well. These techniques and many others are presented in this book. The issue of standardization of the efficiency of solar cells is also discussed.

While many aspects of the performance of organic devices are quite well understood, there are properties that are still a subject of great investigation. As mentioned earlier, one of the open issues is the structural and electronic properties of interfaces between the various organic and inorganic components of the devices. Numerous studies have been performed in order to investigate the electronic structure of the organic/metal electrode and organic/organic interfaces, but they are still not completely understood.[43] The understanding of these interfaces is a central issue for the further development of organic solar cells. Although there are several investigations of the properties and electronic structure of organic/metal electrode and organic/organic interfaces, the direct impact of these interfaces on light harvesting and photocurrent of the organic solar cells has not been completely understood. The electronic structure of the interfaces, in most previous studies, was one of the main factors determining charge injection conditions and therefore device performance. The organic/organic interface has a pronounced impact on the active layer absorption and generation of photocurrent in the visible region of the solar spectrum. This book will discuss new approaches toward engineering new interfaces for better control of the photocurrent generation process.

SPECTROSCOPY OF ORGANIC SOLAR CELL MATERIALS

The use of spectroscopy has been a main tool for the analysis of structure–function relationships important to organic solar cells. A discussion of their use and the results they have provided is of vital importance in grasping a detailed understanding of the materials and science surrounding this field. One such technique is impedance spectroscopy.[44] Impedance spectroscopy has been used for a number of years as a technique providing valuable information on the dark and light behaviors of organic solar cells, among them polymeric-C60 P3HT-PCBM active layers.[45] In this method, a small sinusoidal voltage is superimposed on a bias voltage affecting the space–charge region.[44] The frequency variation of the electrical in-phase and out-of-phase characteristics can show the behavior of the various cell resistances and capacitances found through an equivalent circuit model upon experimental parameters such as bias voltage, chemical treatments, illumination intensity, and duration. Moreover, it permits the determination of several electronic parameters, such as the built-in voltage and the electron-acceptor density together with the width of the depletion region, based on the Mott–Schottky relation, the effective lifetime of the electrons for the recombination process, and the global mobility (Figure 1.6).

In addition to these steady-state measurements, there have been ultrafast techniques such as fluorescence upconversion and transient absorption that have also been used to investigate structure–function relationships in organic solar cells.[46] There has been extensive use of ultrafast fluorescence decay and anisotropy measurements in order to describe intramolecular (and intermolecular) interactions in organic solar materials. In some cases, fluorescence decay is used to probe the process of quenching and resulting exciton diffusion lengths in organic solar materials. In other cases, it is the decay of the fluorescence anisotropy that is used to characterize the interaction strength and energy transfer in the basic repeat units in the case

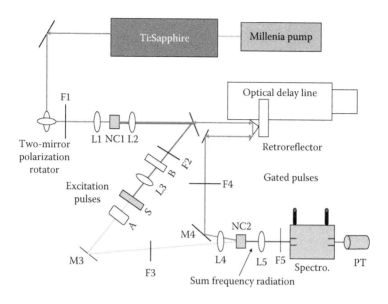

FIGURE 1.6 **(See color insert.)** The basic setup for fluorescence time-resolved measurements.

of polymers and other macromolecular architectures.[47] Femtosecond fluorescence upconversion measurements characterize the dynamics within the first picosecond (<1 ps) of relaxation, which is related to the energy redistribution process. Such processes in particular organic structures have shown an initial fast anisotropy decay profile which requires sophisticated techniques to probe this process. Significantly longer processes are important in organic solar cell materials, but in some cases these processes are related to the rotational motion of the molecular structure and not due to fast energy migration processes which are critical to the description of intramolecular interactions. Investigations of organic solar materials at low (4 K) temperature have also been reported in order to gain further insight into purely electronic processes involved in the dynamics.[49] This is important for discussing the contribution of phonons as it is suggestively suppressed to a great extent. In order to simulate the solar cell geometry, scientists have carried out investigations using this technique by thin films of the active light harvesting molecular structure doped in an inert organic host polymer. Positive results with this approach suggested that the (initial) anisotropy decay time was still fast at lower temperatures. This gave stronger justification that the effect of a fast depolarization at room temperature is indeed related to an electronic mechanism giving rise to a fast energy transfer in the organic material. Also, it was found that the depolarization time decreased with decreasing temperature.[48,50] These measurements, which were very challenging, pushed the limits of our ability to correlate structure–function relationships useful for designing organic solar cells. With these measurements, the field has increased our knowledge about the energy redistribution processes in organic solar materials, which is a focus of this book.

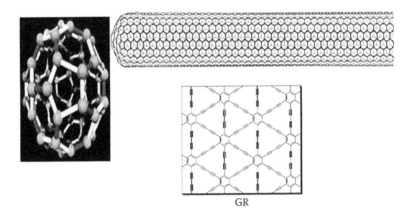

GR

FIGURE 1.7   Molecular structures of C60, carbon nanotube, and two-dimensional carbon network.

Other techniques have also utilized time-resolved methods to investigate the electronic properties of organic solar cell materials. Using transient optical spectroscopy to understand how morphological change enhances device efficiency has attracted significant attention. Time-resolved absorption measurements have also been used to probe the dynamics of excitations in the organic systems which are useful for solar applications (Figure 1.7). For example, transient absorption signals may be analyzed to compare the anisotropy decay obtained from emission (relaxed) with that obtained from direct excitation. Pump-probe ultrafast techniques have been reported extensively in the literature and have been carefully adapted to performing measurements with various molecular systems.[51] Indeed, early studies have been very critical to our understanding of charge regeneration and the charge transfer (CT) states in organic solar cell materials. For example, it has been observed that for some organic systems, the process reported that the CT state in P3HT:PCBM dissociates into charges within a few picoseconds (≈ 8 ps) and recombines to the ground state with a recombination time of ≈ 800 ps. In annealed samples, a smaller fraction of charges recombined in the 800 ps, ≈ 2.5 times fewer than in the unannealed blend, and this was interpreted as the reason for the difference in device efficiency. Other studies employing less intense laser excitation have suggested that the dynamics in the first 10 ps are dominated by diffusion-limited exciton quenching and that the kinetics assigned to charge separation could have been a second order effect.[52] This time scale for exciton diffusion is comparable to that observed in polymer:polymer blends, and investigations to probe this effect are an exciting and challenging part of research in this area. As these effects are not only a result of the chemical structure of the materials but are also connected with their film processing. Indeed, as we will discuss in the interface chapter, some researchers have observed variations in transient absorption results for different film processing methods. For example, researchers have observed that the number of charges remaining 200 ns after photoexcitation in an annealed film was approximately double that in an unannealed film.[53] Investigations of a series of polythiophenes with different ionization potentials and degrees of crystallinity led them to propose that the yield of free charges depends strongly on the degree of crystallinity and on the free energy difference for exciton quenching.[54] It was proposed that a high free energy difference allows fast generation of free charges, with the high excess energy speeding the charge separation.[54] Strong support for an ultrafast mechanism can be obtained by examining the nature of charge recombination before and after annealing. Another major finding suggested that the geminate component of recombination observed before annealing is entirely absent afterwards, indicating that free charges are directly generated as the product of exciton quenching after annealing. The reader will appreciate the fact that the process of film forming and characterization in organic solar cells is an interesting and exciting subject which has ongoing challenges.

New techniques of transient absorption and adaptations of this method have appeared. This is very exciting as some of these methods are reported to detail the transfer of hot electrons from the excited nanostructures to titanium dioxide.[55] Following the high-energy excitation of certain structures such as quantum dots, the hot electrons typically relax to the conduction band-edge within a subpicosecond timescale, losing energy in the process. To extract the electrons before relaxation, this competing process needs to be overcome either by increasing the rate of hot electron extraction or slowing down the cooling. The challenge is, therefore, to create an electron donor–acceptor system that incorporates one or both of these concepts, making the choice of materials vitally important. Other techniques that have focused much of their efforts on carbon nanotubes have applied photoelectrochemical measurements for the possibility of collecting electrons through SnO_2-modified electrodes.[56] As evident from the photocurrent measurements, longer functionalized carbon nanotubes are capable of transporting charges more efficiently than shorter length tubes. A maximum incident photon-to-photocurrent efficiency as high as 32% under an applied potential of 0.2 V vs SCE has been observed with these specialized tubes with this measurement.[56] According to these results, the photoelectrochemical effects of carbon nanotube modified films demonstrate the usefulness of carbon nanotubes in promoting charge transfer and charge transport in photochemical solar cells (Figure 1.8).

The determination of local molecular structure and local atomic organization is often a critical aspect of the analysis of the mobility and lifetime of excitons and charges in the organic solar device. Thus, chemically specific spectroscopic techniques such as Fourier transform infrared (FTIR), solid-state NMR, and UV-vis spectroscopy are extensively used in the study of these materials.[57] X-ray diffraction (XRD) has been used to determine organization on a range of distance scales. In many cases, however, the complexity of the organization requires direct imaging techniques to obtain an adequate understanding of the morphology of a particular material. Scanning electron microscopy (SEM) and transmission electron microscopy (TEM) are by far the most extensively utilized imaging techniques for the characterization of complex materials such as organic solar cell macromolecules.[58] In particular, the use of electron diffraction in combination with electron microscopy is a powerful tool for determining atomic organization and its modulation over longer distance scales. However, certain materials such as organic solar cell materials are less amenable to such analysis. For organic materials, organic–inorganic composites,

FIGURE 1.8   **(See color insert.)** Silicon wafer fabrication and testing facility. (Credit: www. shutterstock.com, 19551091.)

and other materials, atomic force microscopy (AFM) is gaining popularity.[59] This technique is especially useful for determining the topography of the surface of the sample with a resolution on the nanometer, and in favorable cases, angstrom distance scale. While the technique is primarily suited for determining the topography of the sample, certain variations of the technique allow for limited characterization of the chemical structure of the sample through the spatial variation of the tip–sample interaction. For example, regions of the sample that are comprised of soft organic materials can be distinguished from harder inorganic regions of the sample by studying the local compressibility of the sample. Scanning probe tips can also be chemically modified and the spatial variation of the sample tip interaction can be used to infer local chemical composition. Despite advances in developing chemical contrast for AFM, the contrast mechanisms are complex and must be used carefully. There is great need for further development of variations of AFM that offer chemical contrast for a variety of chemical types and morphological situations. Scanning tunneling microscopy (STM) is also a useful technique for characterizing the surface of complex materials, although STM is limited to samples with sufficient conductivity to allow for a detectable electrical current.[60] STM contrast mechanisms are also difficult to control and characterize. Considerable experimental and theoretical advances on the nature of contrast and STM will be necessary in order to more fully apply this technique to emerging molecular materials. Another powerful approach to obtain atomic and molecular specific imaging of materials involves the use of one of the most promising emerging techniques for the characterization of complex materials which is near-field scanning optical microscopy (NSOM).[61] NSOM is a scanning optical microscopy technique that breaks the diffraction limit to ordinary optical microscopy by exciting the sample through a subwavelength aperture in an NSOM probe. A typical NSOM probe is a tapered aluminum-coated single-mode optical fiber that has an aperture at one end of a few tens of nanometers. During the NSOM experiment, the sample's lateral position is raster scanned while the sample probe separation remains fixed. The NSOM tip sample distance is regulated by sensing the interaction force between the tip and the sample. Thus, NSOM leads to simultaneous optical and topographic images of the sample. NSOM offers a correlation of the optical and topographic structure of the samples.[61] Chemical contrast in NSOM is achieved by measuring the local spectroscopic properties of the sample. Usually, fluorescence or UV-vis spectroscopy is studied, although in a few examples, the Raman and infrared vibrational spectroscopy has been spatially resolved with an NSOM probe. The extraordinary sensitivity of fluorescence NSOM allows the observation of NSOM images from sub-monolayers and even single isolated fluorescent molecules. The use of NSOM to study functional organic thin film materials is just beginning. Several recent reviews and well-referenced papers are available.[62] It has already been utilized, however, to explore how the mesostructure of a variety of materials modulates spectroscopic and photophysical properties of these organic solar devices.

The past two decades have witnessed critical technique development in all of the technical areas relevant to the preparation, characterization, and understanding of materials for organic solar cells (Figure 1.9). As a result of this, there are now well-developed and widely applicable tools that can be used to advance the understanding of specific materials systems based on molecular components. The past decade has also brought dramatic progress in the area of price/performance for CPU (central processing unit) cycles and 3D graphics capability, as well as a new generation of computational algorithms. This has led to the ability to carry out detailed

FIGURE 1.9 **(See color insert.)** High-performance computing has played a major role in the discovery of new organic solar materials. (Credit: www.shutterstock.com, 231835942.)

computational studies of remarkably complex molecular systems. Theory, modeling, and simulation incorporate within their domain a wide variety of techniques.[63] These include quantum chemical evaluation of molecular properties and energies, the incorporation of these into semiempirical descriptions, the implementation of such descriptions into molecular mechanics, and the execution of the simulation of molecular assemblies via molecular dynamics and Monte Carlo approaches. Results obtained from such molecularly detailed descriptions can then be folded into more coarse-grained schemes, such as those describing dynamics via continuum characteristics such as friction or those describing dynamics via kinetic schemes. Although a separation of these nonexperimental approaches into the subcategories of theory, modeling, and simulation is often invoked, the interaction among these is a critical aspect of progress, particularly in the context of materials problems. For example, theory based on fundamental quantum mechanical and statistical mechanical formalisms has been critical to the development of completely new algorithms for quantum chemical calculations and for statistical mechanical simulations and have already shown some promise in predicting the properties of materials used in organic solar cells. Also, quantum chemistry is at the heart of results used for the development of intermolecular potentials underlying atomic-scale models, and simulations are being used to develop mechanistic insight that can be incorporated into kinetic models. These are extremely important tools toward the discovery of new materials and device architectures for new organic solar cells.

NEW APPROACHES WITH PEROVSKITES

Finally, a relatively new area of research in organic/inorganic materials has come to realization in the form of perovskites. These metal-containing structures are significantly different from their all organic counterpart; the use of organometallic structures still qualifies them as organic- (perhaps semiorganic-) based devices. The reason for the alarming rate of reports and excitement in this area stems from the fact that there have been reports of solar efficiencies as high as 20% for some of the prepared perovskite-based solar cells. The very high efficiency also comes at a very low cost, mainly due to the relatively simple structure of the organic ligand metal halide structure. New approaches to the fabrication with these structures in TiO_2 and with Al_2O_3 have found that the open-circuit voltage can be impressively large while the binding energy and thermodynamics are minimized.

SUMMARY AND OUTLOOK

Thus, as the introduction to this book comes to a close, it is clear that the area of research for better organic solar cells is both very exciting and very challenging. From a glimpse into this field provided by this book, it is hoped that knowledge of the materials, device fabrication methods, materials characterization, basic physical phenomena, spectroscopy, microscopic investigations, and interfacial properties of organic solar materials will be obtained. It is hoped that from this general summary of this field readers will grasp the excitement and direction the field of organic solar cell research is heading toward. Extensive referencing has been provided in every chapter for readers seeking further details of the topics presented. The literature continues to grow at an impressive pace, and the impact of the work has already made a mark in our understanding of the science and engineering of solar technology. Indeed, the future is bright for organic solar cells, but the road ahead is a tough one. It is suggested that organic solar cells showing 15% efficiency with a 20-year lifetime which can provide electricity at a cost of around seven cents per kilowatt-hour which would make solar energy competitive with conventional sources of electricity. Obviously, the present organic solar devices are a long way from this milestone. However, there is already hope for new materials, new science, and new approaches to this very important problem of creating a solar fuel technology.

REFERENCES

1. Ra. *The Ancient Gods Speak: A Guide to Egyptian Religion*, p. 328, D. B. Redford (ed.), Oxford University Press, New York, 2002.
2. L. Taub, *Ancient Meteorology*, pp. 20–37, Routledge, London, U.K., 2003.
3. A.P. Decandolle, XV11. Biographical memoirs of M. de Saussure, *Philosophical Magazine Series 1*, 4(13), 96–102, 1799.
4. Th J. Seebeck, Magnetische Polarisation der Metalle und Erze Durch Temperatur-Differenz, 1822–23 in Ostwald's Klassiker der Exakten Wissenshaften Nr. 70 (1895).
5. V.V. Vaitheeswaran, *Power to the People: How the Coming Energy Revolution Will Transform an Industry, Change Our Lives, and Maybe Even Save the Planet*, Farrar Straus and Giroux Pubs, New York, 2004.
6. A. Chodos, Bell labs demonstrates the first practical silicon solar cell, *APS News (American Physical Society)*, 18(4), 15, April 2009.
7. R. Williams, Becquerel photovoltaic effect in binary compounds, *The Journal of Chemical Physics*, 32(5), 1960.
8. C. Fritts, On the Fritts selenium cell and batteries, *Van Nostrands Engineering Magazine*, 32, pp. 388–395, 1885.
9. D.M. Chapin, C.S. Fuller, and G.L. Pearson, A new silicon p-n junction photocell for converting solar radiation into electrical power, *Journal of Applied Physics*, 25(676), 1505–1514, 1954.
10. R.S. Ohl, Light-sensitive electric device, US 2402662 A, June 25, 1946.
11. D.M. Chapin, C.S. Fuller, and G.L. Pearson, A new silicon p-n junction photocell for converting solar radiation into electrical power, *Journal of Applied Physics*, 25(5), 676–677, May 1954.
12. H. Son, B. Cartsen, I.H. Jung, and L. Yu, Overcoming efficiency challenges in organic solar cells: Rational development of conjugated polymers, *Energy and Environmental Science*, 5, 8158, 2012.
13. A. Richter, M. Hermle, and S.W. Glunz, Reassessment of the limiting efficiency for crystalline silicon solar cells, *IEEE Journal of Photovoltaics*, 3(4), 1184–1191, October 2013.
14. W. Shockley and W.T. Read, Statistics of the recombinations of holes and electrons, *Physical Review [Internet]*, 87, 835, 1952.
15. J.J.M. Halls and R.H. Friend, Series on photoconversion of solar energy, *Clean Electricity from Photovoltaics*, 1, pp. 377–445, M.D. Archer and R.D. Hill (eds.), Imperial College Press, London, U.K., 2001.
16. W.P. McCray, MBE deserves a place in the history books, *Nature Nanotechnology*, 2(5), 259–261, 2007.

17. H. Hoppe and N.S. Sariciftci, Organic solar cells: An overview, *Journal of Material Research*, 19(7), 1924–1945, 2004.
18. B. Li et al., Review of recent progress in solid-state dye-sensitized solar cells, *Solar Energy Materials and Solar Cells*, 90(5), 549–573, 2006.
19. C.W. Tang, Two-layer organic photovoltaic cell, *Applied Physics Letters*, 183, 48, 1996.
20. P. Peumans, A. Yakimov, and S.R. Forrest, Small molecular weight organic thin-film photodetectors and solar cells, *Journal of Applied Physics*, 93, 3692, 2003.
21. P. Peumans, S. Uchida, and S.R. Forrest, Efficient bulk heterojunction photovoltaic cells using Small molecular weight organic thin films, *Nature*, 425, 158–162, 2003.
22. S.J. Park, J.M. Cho, W.-B. Byun, J.-C. Lee, W.S. Shin, I.-N. Kang, S.-J. Moon, and S.K. Lee, Bulk heterojunction polymer solar cells based on binary and ternary blend systems, *Journal of Polymer Science Part A: Polymer Chemistry*, 49(20), 4416–4424, 2011.
23. W. Wilhem, Polymer film and laminate technology for low-cost solar energy collectors, *Polymers in Solar Energy Utilization*, Chapter 3, pp. 27–38, Chapter DOI: 10.1021/bk-1983-0220.ch003, ACS Symposium Series, American Chemical Society, Vol. 220.
24. M. Grätzel, Solar energy conversion by dye-sensitized photovoltaic cells, *Inorganic Chemistry*, 44(20), 6841–6851, 2005.
25. B. O'Regan and M. Grätzel, A low-cost high-efficiency solar cell based on dye-sensitized colloidal TiO$_2$ films, *Nature*, 335, 737, 1991.
26. W. West, Proceedings of vogel centennial symposium, *Photographic Science and Engineering*, pp. 18–35, 1974.
27. P. Heremans, D. Cheyns, and B. Rand, *Accounts of Chemical Research*, 42(11), 1740–1747, 2009.
28. K. Lobato, L.M. Peter, and U. Würfel, Direct measurement of the internal electron quasi-Fermi level in dye sensitized solar cells using a titanium secondary electrode, *Journal of Physical Chemistry B*, 110(33), 16201–16204, August 24, 2006.
29. U.S. Department of Energy, Photovoltaic cell conversion efficiency basics. http://energy.gov/eere/energybasics/articles/photovoltaic-cell-conversion-efficiency-basics. Accessed August 16, 2013.
30. B. Qiab and J. Wang, Fill factor in organic solar cells, *Physical Chemistry Chemical Physics*, 15, 8972–8982, 2013.
31. X. Guo et al., *Nature Photonics*, 7, 825–833, 2013.
32. A.C. Mayer, S.R. Scully, B.E. Hardin, M.W. Rowell, and M.D. McGehee, Polymer based solar cells, *Materials Today*, 10(11), 28, 2007.
33. H. Najafov, B. Lee, Q. Zhou, L.C. Feldman, and V. Podzorov, Observation of long-range exciton diffusion in highly ordered organic semiconductors, *Nature Materials*, 9(11), 938, 2010.
34. D.C. Coffey, A.J. Ferguson, N. Kopidakis, and G. Rumbles, Photovoltaic charge generation in organic semiconductors based on long-range energy transfer, *ACS Nano*, 4(9), 5437–5445, September 2010.
35. J. Xue, Perspectives on organic photovoltaics, *Polymer Reviews*, 50(4), 411, 2010.
36. R. Steim, The impact of interfaces on the performance of organic photovoltaic cells, DOKTOR INGENIEURS, Karlsruher Institut für Technologie (KIT), KIT Scientific Publishing, Karlsruhe, Germany, 2010.
37. R.M. Dickson, A.B. Cubitt, R.Y. Tsien, and W.E. Moerner, On/off blinking and switching behaviour of single molecules of green fluorescent protein, *Nature*, 388, 355–358, 1997.
38. G. Binnig, C.F. Quate, Ch. Gerber, Atomic force microscope, *Physical Review Letters*, 56(9): 930–933, 1986.
39. D.C. Coffee and D.S. Ginger, Time-resolved electrostatic force microscopy of polymer solar cells, *Nature Materials*, 5, 735–740, 2006.

40. M. Nonnenmacher, M.P. O'Boyle, and H.K. Wickramasinghe, Kelvin probe force microscopy, *Applied Physical Letters*, 58(25), 2921, 1991.
41. W. Melitz, J. Shen, A.C. Kummel, and S. Lee, Kelvin probe force microscopy and its application, *Surface Science Reports*, 66, 1–27, 2011.
42. T. Ishida, W. Mizutani, Y. Aya, H. Ogiso, S. Sasaki, and H. Tokumoto, Electrical conduction of conjugated molecular SAMs studied by conductive atomic force microscopy, *Journal of Physical Chemistry B*, 106(23), 5886–5892, 2002.
43. C. Poelking, M. Tietze, C. Elschner, S. Olthof, D. Hertel, B. Baumeier, F. Würthner, K. Meerholz, K. Leo, and D. Andrienko, Impact of mesoscale order on open-circuit voltage in organic solar cells, *Nature Materials*, 2014, DOI: 10.1038/nmat4167.
44. G. Garcia-Belmonte, A. Munar, E.M. Barea, J. Bisquert, I. Ugarte, and R. Pacios, Charge carrier mobility and lifetime of organic bulk heterojunctions analyzed by impedance spectroscopy, *Organic Electronics*, 9, 847–851, 2008.
45. P. Schilinsky, C. Waldauf, and C.J. Brabec, Recombination and loss analysis in polythiophene based bulk heterojunction photodetectors, *Applied Physics Letters*, 81, 3885, 2002.
46. T. Goodson III, Optical excitations in organic dendrimers investigated by time-resolved and nonlinear optical spectroscopy, *Accounts of Chemical Research*, 38(2), 99–107, 2005.
47. O.P. Varnavski, J.C. Ostrowski, L. Sukhomlinova, R.J. Twieg, G.C. Bazan, and T. Goodson III, Coherent effects in energy transport in model dendritic structures investigated by ultrafast fluorescence anisotropy spectroscopy, *Journal of the American Chemical Society*, 124(8), 1736–1743, 2002.
48. O. Varnavski, G. Menkir, T. Goodson III, P.L. Burn, I.D.W. Samuel, J.M. Lupton, and R. Beavington, Ultrafast polarized fluorescence dynamics in an organic dendrimer, *Applied Physics Letters*, 78, 3749, 2001.
49. M. Ranasinghe, Y. Wang, and T. Goodson III, Excitation energy transfer in branched dendritic macromolecules at low (4 K) temperatures, *Journal of the American Chemical Society*, 125(18), 5258–5259, 2003.
50. H. Ohkitaa and S. Ito, Transient absorption spectroscopy of polymer-based thin-film solar cells, *Polymer*, 52(20), 4397.
51. J. Cabanillas-Gonzalez, G. Grancini, and G. Lanzani, Pump-probe spectroscopy in organic semiconductors: Monitoring fundamental processes of relevance in optoelectronics, *Advanced Materials*, 23(46), 5468, 2011.
52. H. Ohkita and S. Ito, Exciton and charge dynamics in polymer solar cells studied by transient absorption spectroscopy, *Organic Solar Cells*, W.C.H. Choy (ed.), Green Energy and Technology, Springer-Verlag, London, U.K., 2013.
53. Y. Kim, S.A. Choulis, J. Nelson, D.D.C. Bradley, S. Cook, and J.R. Durrant, Device annealing effect in organic solar cells with blends of regioregular poly(3-hexylthiophene) and soluble fullerene, *Applied Physics Letters*, 86, 063502, 2005.
54. N. Bansal et al., Influence of crystallinity and energetics on charge separation in polymer–inorganic nanocomposite films for solar cells, *Nature*, 3, 1531, 2013.
55. X. Jin, Q. Li, Y. Li, Z. Chen, T.-H. Wei, X. He, and W. Sun, Energy level control: Toward an efficient hot electron transport, *Nature*, 4, 5983, 2014.
56. A. Furuta, T. Kuramoto, and T. Arai, Additional layer of the multi-walled carbon-nanotubes increases the photo-current of a poly(3-hexylthiophene)-sensitised solar cell, *Energy Environmental Science*, 2, 853–856, 2009.
57. R.D. Pensack, C. Guo, K. Vakhshouri, E.D. Gomez, and J.B. Asbury, Influence of acceptor structure on barriers to charge separation in organic photovoltaic materials, *Journal of Physical Chemistry C*, 116, 4824–4831, 2012.
58. B.G. Mendis and K. Durose, Prospects for electron microscopy characterisation of solar cells: Opportunities and challenges, *Ultramicroscopy*, 2011, DOI: 10.1016/j.ultramic.2011.09.010.

59. D. Mark, J. Peet, and T.-Q. Nguyen, Nanoscale charge transport and internal structure of bulk heterojunction conjugated polymer/fullerene solar cells by scanning probe microscopy, *Journal of Physical Chemistry C*, 112, 7241, 2008.

60. X.-D. Dang, A.B. Tamayo, J. Seo, C.V. Hoven, B. Walker, T.-Q. Nguyen, *Advanced Functional Materials*, 20(19), 3314, 2010.

61. C.R. McNeill, H. Frohne, J.L. Holdsworth, J.E. Furst, B.V. King, and P.C. Dastoor, Direct photocurrent mapping of organic solar cells using a near-field scanning optical microscope, *Nano Letters*, 4(2), 219–223, 2004.

62. H. Zhou, A. Midha, G. Mills, L. Donaldson, and J.M.R. Weaver, Scanning near-field optical spectroscopy and imaging using nanofabricated probes, *Applied Physics Letters*, 75, 1824–1826, 1999.

63. V. Coropceanu, H. Li, P. Winget, L. Zhu, and J.-L. Brédas, Electronic-structure theory of organic semiconductors: charge-transport parameters and metal/organic interfaces, *Annual Review of Materials Research*, 43, 63–87, 2013.

2 New Materials for Solar Energy Conversion

The need for superior materials for solar devices has enjoyed a new sense of enthusiasm. Indeed, the organic materials to be used in solar cells have been developed substantially in the last 30 years. It wasn't long ago that many of the organic materials suggested for solar cells were thought to be inefficient, terribly unstable, and too expensive to ever be considered for mass production or translation and commercialization (Figure 2.1).[1] Some of these problems have, to a great extent, been reduced with new ideas and approaches toward chemically modifying certain organic small molecule and polymeric structures. In this chapter, a major goal is to understand the basic steps in the development that has led us to modern organic materials, which are now showing great promise. From a chemical standpoint, it is the process of developing organic solar materials that has been fascinating. It's not often that synthetic chemists, physical chemists and physicists, and engineers have worked so closely together on solving an important cross-disciplinary problem.

SMALL MOLECULES FOR ORGANIC SOLAR CELLS

The synthesis and device fabrication of organic solar cell materials can be grouped into two categories. The first category concerns the synthesis of small molecular systems with high purity. The second group of organic solar materials focuses on the creation of conjugated polymers and other macromolecular architectures with wide bandgaps. In terms of their practical use, small organic molecules for solar applications provide good flexibility in using synthetic organic chemistry to modify the basic structure–function relationships of the particular application.[2] Many of these molecular structures have a history of previous use in other optical and electronic applications, such as light-emitting diodes (LEDs),[3] nonlinear optics (NLO),[4] and molecular sensing.[5] The reader can easily see the similarity with these other optical and electronic effects. Many of the principles for the creation and use in these applications hold true for solar devices as well. For example, in organic small molecular structures, a high efficiency in absorption and subsequent emission is desired in the visible and near infrared (IR) spectral wavelengths.[6] In this approach, organic thiophenes[7] as well as organometallic phosphor materials[8] have been prepared in the past to cover these regions. Particular care is taken to investigate the functionality and metal–ligand interactions to enhance the effects in fabricated materials and devices for LED and later solar applications.[6] As it was already known in the case of organic nonlinear optical materials,[4,9] the need for relatively large transition dipole moments and spectral absorption in the visible to near-IR regions as well as good energy transfer properties led to the use of these materials

FIGURE 2.1 **(See color insert.)** Fabrication of flexible solar cells.

in a variety of electronic applications. It was found that two particular organic molecules were useful in this situation. Both porphyrin[10] and pthalocyanine[11] systems showed the right combination of transition dipole moment and spectral distribution to be used for organic optoelectronic effects. The information obtained from the wealth of studies in this direction was later directed to photovoltaic applications. Various functionalized porphyrin[12] and pthalocyanine[13] systems have been prepared and tested for photovoltaic applications. This was the initial point for the synthesis of organic small molecule solar cell materials. Scientists and engineers have explored a wide range of small molecular systems through a tour-de-force use of synthetic organic chemistry. In this chapter, we explore a sampling of these wonderful organic small molecular systems and devices and their impact on the direction of the field of organic solar devices.

Perhaps the best way to discuss the subject of organic materials for solar cells is to begin with a very popular class of organic solar cells (OSCs), the bulk heterojunction (BHJ) solar cells, composed of low-bandgap, conjugated polymers as electron donors, and substituted fullerenes as the electron acceptor.[14] To date, there has been remarkable progress in the performances of these cells, resulting from successive developments in material design and synthesis, control in the morphology of the BHJ composite[16] and device optimization.[17] Therefore, power conversion efficiencies of more than 8% have been achieved.[18] Efficiencies approaching 10% were disclosed for a small area solar cell.[10] This indicated the rapid progress over the last 10 years and the great potential of polymer solar cells as an alternative source of energy. In order for organic solar cells to fully mature from research and development into cost-effective products, continuous improvement in solar cell efficiency must be achieved. A Power conversion efficiency (PCE) of 10% or more in devices with sizable area is regarded as an important threshold for practical and widespread usages of polymer solar cells. It is reported that organic solar cells showing 15% PCE with a 20-year lifetime can provide electricity at a cost of around seven cents per kilowatt-hour, which would make solar energy competitive with conventional sources of electricity.[20] In order to achieve this PCE goal, major advances in new materials and device technology are critical. A fundamental comprehension of the mechanisms occurring in organic solar cells and an understanding of design principles of new materials is necessary to push this area forward (Figure 2.2).

It should be made clear from the start that bulk heterojunction solar cell devices are sought as alternatives for silicon-based devices.[21] The issue is the relatively low efficiencies in these systems. The strong motivation for approaching such efficiencies is driving various research efforts in the area of organic-based BHJ solar cells.[22] Molecular crystalline semiconductors as alternatives to conjugated polymers offer several intrinsic advantages in solution processable BHJ solar cells.[21] Owing to

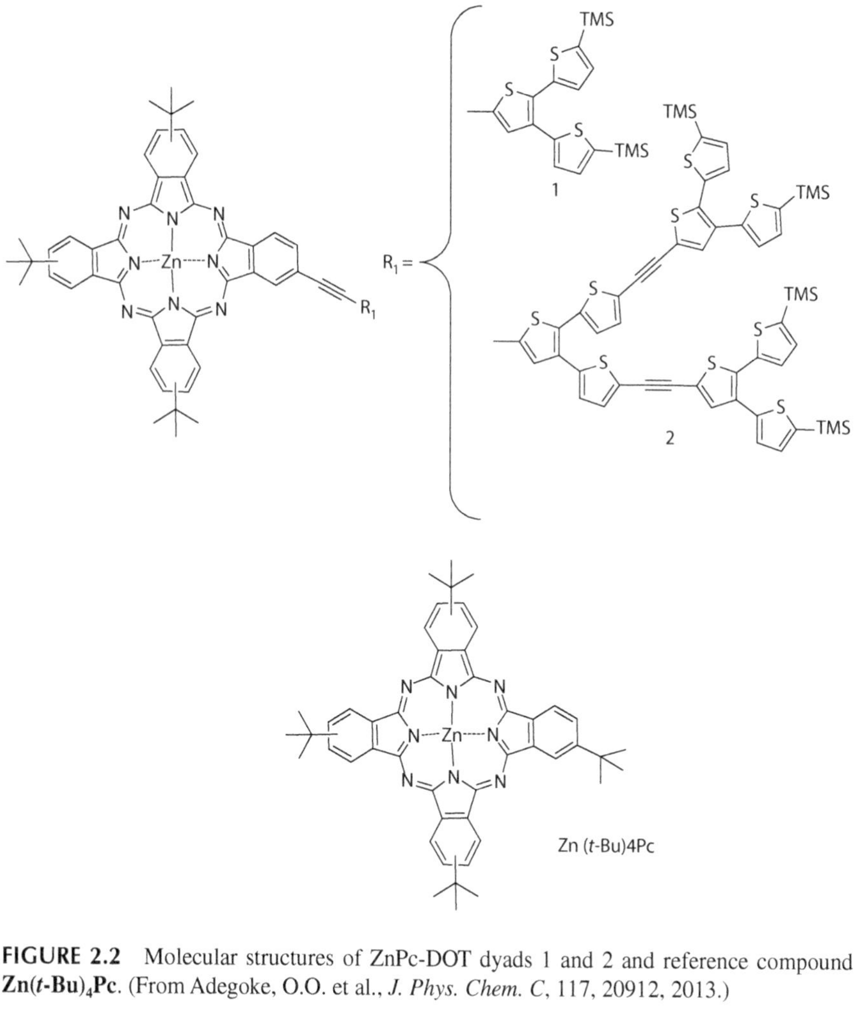

FIGURE 2.2 Molecular structures of ZnPc-DOT dyads 1 and 2 and reference compound **Zn(t-Bu)₄Pc**. (From Adegoke, O.O. et al., *J. Phys. Chem. C*, 117, 20912, 2013.)

their monodisperse nature with well-defined chemical structures, together with the absence of end-group contaminants or batch-to-batch variations, molecular organic semiconductors can be reproducibly prepared, functionalized, and purified.[23] Despite the fact that the highest efficiencies to date for small molecule-based solar cells remain lower than their polymer-based analogs, the considerations mentioned earlier make molecular crystalline semiconductors attractive as active materials in BHJ solar cells.[24] Encouraging reports of diketopyrrolopyrrole oligothiophenes as donor materials in molecular BHJ solar cells have shown PCEs of 2.2%–4.4%.[25] Isoindigo-based oligothiophenes have been used as donor materials in molecular BHJ solar cells and also in conjunction with bithiophene as an electron donor.[26] Devices constructed with these materials exhibit very promising power conversion efficiencies. This is a promising approach for solution-processed small molecule solar cells. Synthetic chemists have shown great expertise in providing combinations of these donor–acceptor systems to probe the extent of charge transfer and exciton generation in BHJ structures. In some cases, di-block (and multi-block) copolymers have been prepared (Figure 2.3).

In general, organic semiconductors can be regarded as versatile electronic materials. Their properties can vary from intrinsic wide-bandgap semiconductors (bandgaps above 1.4 eV) down to insulators (bandgaps above 3 eV) with a negligibly low intrinsic charge carrier density at room temperature in the dark. Materials having

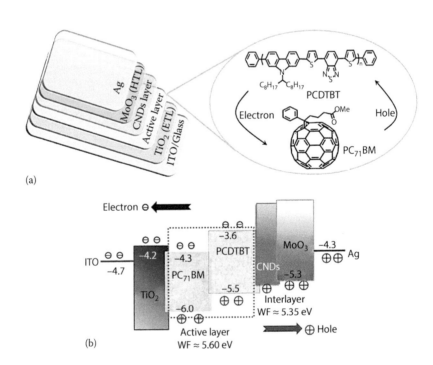

FIGURE 2.3 (a) Device structure of the inverted BHJ-OSCs. (b) Scheme of energy levels of the materials involved in inverted polymer solar cells. (From Zhang, X. et al., *J. Phys. Chem. C*, 120, 13954, 2016.)

a delocalized π electron system can absorb sunlight, create photogenerated charge carriers, and transport these charge carriers.[28–30] Research with conjugated organic solar cells generally focuses on either solution processable organic semiconducting small molecules or on vacuum-deposited small-molecular materials. As an example, phthalocyanine and perylene have commonly found applications in thin film organic solar cells.[31,32] Phthalocyanine is a p-type[33] hole-conducting material that works as an electron donor. However, perylene and its derivatives show an n-type, electron conducting behavior.[34] Together, these two materials usually serve as electron-acceptor materials. Chemical, photochemical, or electrochemical doping is used to introduce extrinsic charge carriers into organic semiconductors.[35–37] A very common process of photoinduced electron transfer from a donor to an acceptor-type organic semiconductor film introduces free charge carriers in the device. Donor-acceptor-type bilayer devices can thus work like larger macromolecular systems but there is more control as the small molecule chemistry can be manipulated and the final products accurately characterized.

A number of important research investigations have considered smaller donor acceptors, such as tetracene as the active materials in BHJ solar cells.[38] In combination with C_{60}, various test devices have been constructed and the electronic properties of these systems have been probed. For example, the UV–Vis absorption spectrum of tetracene shows a peak absorption near the middle of the visible spectrum with a maximum near 520 nm and an absorption edge at 580 nm. On the other hand, C_{60} contributes a significant absorption in the short-wavelength range with an absorption maximum at 440 nm.[39] Although the absorption spectra of tetracene and C_{60} films overlap in the short-wavelength range, both the tetracene and the C_{60} layers contribute to the photogeneration of the carriers and show complementary absorption spectra in the visible range.[40] By optimizing the growth rate of tetracene films, researchers have worked to improve the performance of small molecular devices. It has been found that these tetracene-related systems hold promise of efficient, low cost, large area "plastic solar cells."[41] A major advantage of using small molecules is the relative high purity and control of film thickness and crystallinity. Molecules listed in Figure 2.4 are just a sample of some of the frequently used small molecules researchers have investigated to create organic solar cells.

While a number of different small molecule systems have been tested, only those which illustrate processes important to controlling the operation of solar cells have been thoroughly studied. Presently, a key limitation in organic photovoltaic devices

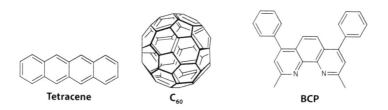

Tetracene C_{60} **BCP**

FIGURE 2.4   The molecular structure of the tetracene and C_{60} and BCP chromophores.

is their low open circuit voltages (V_{oc}), which is typically less than half of the incident photon energy.[42] There is an important structure–function relationship between the materials' properties and the open circuit voltage. The power conversion efficiency (η) of the solar cells was calculated from the following equation:

$$\eta = \frac{FF * J_{sc} * V_{oc}}{P_{in}} 100\% \qquad (2.1)$$

where
  FF is the fill factor
  J_{sc} is the short circuit current density
  V_{oc} is the open circuit voltage
  P_{in} is the power density of the incident light

Researchers have systematically investigated the basic processes that control and limit the open-cell voltage in organic solar cells. This can serve as a guide for the identification of new materials for high-performance (high-efficiency) devices.[43] Important detailed studies have been carried out to probe this correlation with structure and open-cell voltage. For example, to determine the molecular structure–property correlations with the open-cell voltage for a given current, researchers have considered the previously mentioned donors, tetracene and rubrene.[44] Despite the similarity of the highest occupied molecular orbital (HOMO) energies of 5.1 eV for tetracene and 5.3 eV for rubrene, the open-cell voltage (V_{oc}) for the two cells are considerably different at 0.55 and 0.92 V, respectively.[44] The difference in V_{oc} is due to the large difference in the current dependent process. The J_{sc} is two orders of magnitude larger for tetracene than for rubrene. It is well known that tetracene is a planar molecule consisting of fused conjugated aromatic rings.[45] In contrast to this, the orthogonal pendant phenyl rings of rubrene prevent close association of the tetracene cores of adjacent rubrene molecules.[46] This result is clearly seen from comparisons between the solution and thin film UV–Vis absorption spectra of the two compounds.[47] While rubrene gives similar solution and thin film spectra, strong intermolecular π interactions in tetracene thin films yields a broadened and red-shifted spectrum relative to the solution. Weak intermolecular interactions in rubrene decrease both the donor–donor as well as the donor–acceptor interactions relative to tetracene. It is this kind of molecular interpretation that drives the rational thinking for better organic molecules for solar applications. The collaboration between scientists working on the synthesis of new small molecules and researchers doing physical measurements to test the structure–function correlations is a critical part of the discovery process for new organic solar cell materials. The small molecule polyene-type structures were a good start as they could be made in high yield and high purity and one could alter the pi-structure strongly with synthetic control. In addition to polyenes, several initial reports were devoted to illustrating the interesting V_{oc} properties measured for copper phthalocyanine (CuPc)- and platinum tetraphenylbenzoporphyrin (PtTPBP)-containing devices as well. It was shown that there can be a structure–function correlation of these systems with the device operation connected to intermolecular interactions.[48] It is interesting to note how two similar compounds have comparable π-systems but vary strongly in their V_{oc}. While a larger V_{oc} would be expected for the CuPc-based device, the experimental CuPc/C$_{60}$ V_{oc} is actually less than that of PtTPBP by 0.21 V.[48,49] Here, again, a molecular rationalization of this result is used. The molecular interpretation is that CuPc is a flat molecule with strong intermolecular interactions in the solid state. In contrast, PtTPBP has a highly distorted saddle-shaped conformation with four orthogonal phenyl rings that reduce the availability of the π-system to intermolecular interactions and electronic delocalization.[49] Indeed, structure (both electronic and conformational) plays a major role in the V_{oc} and this is what the scientists have utilized as the vehicle to drive toward better solar organic materials.

A very popular approach has been to further tailor the properties of organic small molecules by synthetic procedures to adjust their electronic resonance. For example, in understanding the effect with CuPc, which shows a large broadening of its low-energy absorption in its thin film UV–Vis spectra, while the solution and thin film spectra of PtTPBP (see Figure 2.5 structures) are the same line width, researchers have modified these molecules synthetically to probe this effect.[50] It has been shown that additional steric bulk, by synthetically appending phenyl rings, decreases the degree of intermolecular interaction for PtTPBP relative to CuPc.[50] The high V_{oc} for the PtTPBP device results from diminished recombination and hence a lower J_{sc} than for the CuPc device. PdTPBP has been examined as a donor material as well.[51] The structure of the Pd analog is close to that of PtTPBP, where the Pd substitution shifts the HOMO energy. However, in this case it was suggested that this does not affect the strengths of the intermolecular interactions. This results in a correlation that for Pd and Pt complexes there is a very small difference in the J_{SO}.[52] Electronic interactions, or coupling between donor and acceptor molecules, can be modified by varying the strength of the π-conjugation. This is a concept long appreciated by organic chemists. Researchers have prepared Pt tetraphenyl naphtholporphyrin (PtTPNP), an analog of PtTPBP whose π-system is extended by the benzanulation of four rings onto the benzo moieties of PtTPBP.[49,51] The differences in properties in terms of the V_{oc} could be explained by changes in conformational energy studies.[52] The lowest energy conformation of the naphthol analog has a saddle shape similar to PtTPBP, resulting from steric repulsions of the meso and pyrolle substituents.[53]

The area of small organic molecule solar devices continues to grow. Novel materials containing organometallic materials as well as novel charge transfer chromophores have been employed and continue to be optimized. As mentioned earlier, it must be remembered that the approach of correlating the structure–function relationships with particular device parameters has proven to be very valuable. While there can be some success based on multiple adaptations of particular structures, scientists in this area strive to develop good molecular correlations with the possibility of predicting the properties of a new device material. This process continues for small molecule systems and, in many cases, for larger macromolecule systems for the case of conjugated polymers.

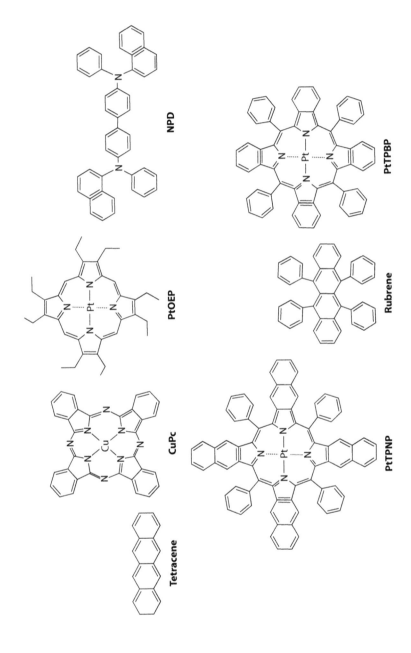

FIGURE 2.5 The molecular structure of rubrene and related chromophores.

ORGANIC CONJUGATED POLYMER FOR SOLAR CELLS

Organic conjugated polymers have played a major role in the electronic applications of semiconductors.[54–56] As mentioned earlier, structures such as thiophenes, alkenes, and phenylenevinylene systems have shown excellent LED, PV, and NLO properties. In consideration of making new solar materials, the advantages polymers have stemmed from their processing properties, stability, and possibly lower cost. In nearly every conjugated polymer, inert alkyl side chains are introduced to achieve good processibility.[57] Ideally, the alkyl side chains should be employed in such a manner as not to disturb the closely packed aromatic backbone structures while maintaining adequate solubility for solution processing. Researchers have found that the charge carrier mobility can be optimized due to this effect. The polymer's backbone planarity and thermodynamically driven self-organization in the solid state are connected with this process. However, many polymers have a high tendency to crystallize, which can reduce polymer chain interaction with acceptor molecules.[58] One observes, in this case, phase separation and therefore less than ideal morphology in the polymer BHJ solar cells. Researchers agree now that it is important to synthesize side chains that will improve interactions without disturbing the polymer's crystallinity in the film. It has been found that polymer solar cells showed very different characteristics when incorporating different alkyl side chains on the donor polymers.[59] This is most likely due to changes in the film morphology resulting from different conformations of alkyl side chains. As we will see in Chapter 5 on interfaces, the nature of the polymer backbone can have a significant influence on the morphology of a polymer blend film and, therefore, the performance of the polymer solar cell.[60] As the reader can see from this brief discussion, the use of appropriate conjugated polymers for solar applications is not trivial. There are a number of factors (in some cases competing) that can either help or limit the performance of the solar device when using conjugated polymers.

As in many applications using organic polymers, the solution and film-forming properties are important considerations. In general, when thinking about real devices, polymers with molecular properties inducing unfavorable interactions with PCBM (C_{60}) molecules form immiscible blend films with large-scale phase segregation. For example, perfluorination of the donor polymer's backbone typically makes the polymer less interactive and so, less miscible with PCBMs by inducing the fluorophobic effect.[61] This effect gave a strong driving force for phase separation, leading to the formation of domains on the 100 nm scale.[61] Unfortunately, this resulted in a weak solar cell performance. The synthesis of good miscible conjugated polymers became a major goal in the research of BHJ solar cells. Another factor that was also considered in conjunction with the miscibility was the surface energy. The surface energy in a polymer film is defined as a dominant force for spreading a liquid across the surface. Researchers have found that the surface energy affects the conjugated polymer's morphology. For the case of organic conjugated polymers that have surface energies similar to C_{60}, it was found that highly miscible blend films are formed, which have small or nanoscale phase segregation morphology. This is opposite to the case of polymers with larger surface energy differences to PCBM, which caused large domain sizes and phase separation.[63] Therefore, the blends with closely matched surface energies showed the highest solar cell performance. This has helped guide the research design of potential organic polymer solar cell devices.

In general, conventional single-layer Schottky barrier and two-layer heterojunction photovoltaic cells from conjugated polymers generally have poor photon-to-current conversion efficiencies. There have been modest improvements by the use of acceptor systems.[65] Because most conjugated polymers that have been used in photovoltaic devices are *p*-type semiconductors, a major obstacle to improving these devices is the scarcity of suitable *n*-type polymers.[66] However, it should be stated that the two-layer *p/n* ~ donor/acceptor heterojunction is an ideal polymer solar cell architecture if the *p*- and *n*-type layers exhibit complementary optical and electrical properties.[67] This works best when there is an ohmic contact at the interface. The poor performance of some of the bilayer polymer devices (with efficiencies of only ~0.5%) has been attributed to the limitations of exciton diffusion, to the interface where charge separation occurs, and to carrier transport to the collecting electrodes. The ability to make ohmic contacts to *n*-type conjugated polymers is very difficult because of the requirement of low work function metals, which are unstable in air, to match the electron affinity of the polymer.

As mentioned in Chapter 1, researchers in the area of OLEDs have applied their knowledge concerning conjugated polymers and film/device formation to different areas of research. For example, a system composed of poly~*p*-phenylene vinylene (PPV) and the *n*-type conjugated ladder polymer poly~benzimidazobenzophenanthr oline ladder (~BBL), which is similar to a well-known OLED material, was reported for solar applications.[68] The researchers investigated PPV/BBL heterojunctions because of the complementary electronic structures and optical properties of the conjugated polymers. These two polymers seem to work well together. The PPV thin films have good *p*-type semiconducting properties that are well known from the vast amount of research done on OLEDs with them.[69] BBL was shown to have promising *n*-type semiconducting, photoconductive, and nonlinear optical properties.[68] And the absorption width of PPV covers 300–520 nm, while the BBL absorption extends to the near infrared (~740 nm), suggesting great potential for light harvesting in the PPV/BBL bilayers.[68] This is a good example of research that has crossed into different fields. The same electronic properties of these novel materials are good for both OLEDs and solar applications (Figure 2.6).

The difficult part in developing conjugated BHJ solar cells is to create a particular strategic design for success in terms of the materials and bandgap. After a number of initial attempts, the field has localized on a particular strategy that seems fairly consistent with the better performing materials and devices. Generally, this strategy includes (1) reducing the bandgap of polymers so as to harvest more sunlight, which leads to a higher short-circuit current density (J_{sc}),[70] and (2) lowering the highest occupied molecular orbital (HOMO) of the polymers, which increases the open circuit voltage (V_{oc}).[71] With the rise in interest in using low-bandgap polymers to harvest more sunlight from longer wavelengths, much effort has been made recently in reducing the bandgap of polymers.[70,72] Some scientists have reported organic polymers with efficiencies of over 5% using different low-bandgap polymers.[72] The extended absorption of sunlight at longer wavelengths directly reflects on the value of J_{sc}. A relatively high current density has been achieved.[70,73] On the other hand, organic

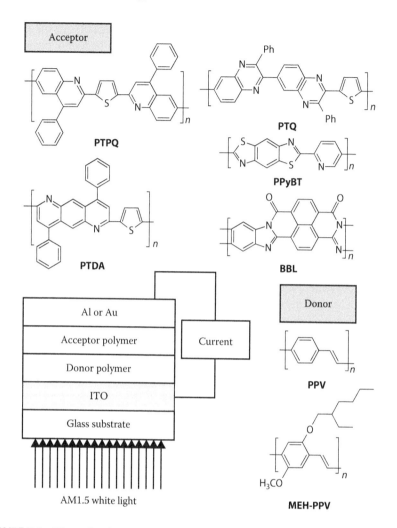

FIGURE 2.6 The molecular structure of MEH-PPV and BBL and the device structure.

conjugated polymer solar cells with a high V_{oc} have been realized by other research groups using polymers that absorb at shorter wavelengths.[74] To push the efficiencies toward the predicted theoretical limit, however, achieving both a high J_{sc} and a high V_{oc} is essential. To match the energy level of the commonly used electron acceptor C_{60} (PCBM), both the HOMO and the lowest unoccupied molecular orbital (LUMO) of the polymer need to be considered while tuning the bandgap of the conjugated polymer. It is known that the energy difference between the LUMOs of donor and acceptor should be larger than 0.3 eV for efficient charge separation, which directly relates to the J_{sc} of solar cells.[75] However, the V_{oc} of polymer cells is limited by the difference between the HOMO of the donor and the LUMO of the acceptor. As a result, narrowing the bandgap of polymers without sacrificing efficient charge separation as well as high V_{oc} becomes a major hurdle in achieving high efficiency. To address this, scientists have tried to optimize the donor and acceptor properties of the polymers as described earlier. In some cases, an additive effect could be found. For example, researchers have altered the HOMO of poly[4,8-bis-substituted-benzo [1,2-b:4,5-b0] dithiophene-2,6-diyl-alt-4-substituted-thieno[3,4-b]thiophene-2,6-diyl] (PBDTTT)-derived polymers by adding different electron-withdrawing functional groups, step by step.[76] It was found that the addition of more than one electron-withdrawing group is effective in further lowering the HOMO of PBDTTT.[70,76] This result was very significant. It suggests that, in general, tuning the V_{oc} of conjugated polymer solar cells by means of molecular design may be realized using a step-by-step approach. Applying stronger electron-withdrawing groups to the backbone of polymers has been found to be effective in lowering the HOMO of polymers, which directly affects the V_{oc} of the organic polymer solar cell. Based on this approach for the PBDTTT polymer derivative system, an efficiency higher than 7% was achieved by combining the advantages of a low HOMO level in the polymer (high V_{oc}) and long wavelength absorption (high J_{sc}). There is still a great deal of room for improvement in increasing the V_{oc} by tuning the energy levels of the polymers to match the energy levels of the PCBM. A further improvement in efficiency can be expected if the HOMO of the polymer can be further lowered without loss of J_{sc} (Figure 2.7).[77]

While much focus is put on the bandgaps of the conjugated polymers, there is also the possibility of tuning the fullerene bandgap in connection with the energy levels of the polymer. Researchers have found a series of highly soluble fullerene derivatives with varying acceptor strengths (first reduction potentials), which can be synthesized and used as electron acceptors. These fullerene derivatives showed a variation of almost 200 mV in their first reduction potential.[78] The open circuit voltage of the corresponding devices was found to correlate directly with the acceptor strength of the fullerenes, whereas it was rather insensitive to variations of the work function of the negative electrode. These observations are discussed within the concept of Fermi level pinning between fullerenes and metals via surface charges.

An even more pronounced effect was found for the change in V_{oc} with the change in fullerene reduction potential. For example, the highest and the lowest average open circuit voltages were observed for PCBM- and ketolactam (see Figure 2.8 for structures)-containing cells, with 760 and 560 mV, respectively. It was shown that the open circuit voltage in these polymeric solar cells is directly related to the acceptor strength of the fullerenes. This result fully supports the view that the open circuit voltage of this type of donor-acceptor bulk heterojunction cell is related directly to the energy difference between the HOMO level of the donor and the LUMO level of the acceptor components. These observations led the way to molecular engineering of the open circuit potentials in organic polymer solar cells by tailoring the electronic structure of the acceptor (Figure 2.9).

In addition to small molecules and linear conjugated polymers, some researchers have instituted the use of copolymers in BHJ organic solar cells. For example, some have made a series of donor copolymers incorporating various polymers with benzo[1,2-b;4,5-b']dithiophene (BDT) (PTBs). The planarity of the BDT with an extended π- system conveys rigidity to the polymer backbone, enabling the polymer to form an assembly with better π–π stacking and higher hole mobility. It was shown that these polymers have components that can stabilize the quinoidal structure of

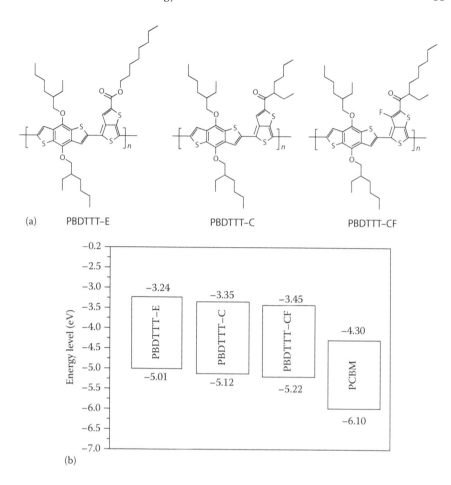

(a) PBDTTT–E

PBDTTT–C

PBDTTT–CF

(b)

FIGURE 2.7 (a) The molecular structure of PPV and BBL and (b) the energy level diagram.

the polymer, which contributes to the lowering of the bandgap with efficient absorption (around 700 nm) in the region of greatest photon flux of the solar spectrum. The ester electron-withdrawing group in the polymer plays a role in lowering the HOMO energy levels, stabilizing the monomer against oxidation and increasing solubility. The PTB polymer showed very promising properties for OPV applications with a maximum of 690 nm in the optical absorption spectrum and good miscibility with fullerene acceptors.[70,80] From grazing incidence wide-angle x-ray scattering (GIWAXS) studies, the polymer chains were found to be stacked on the substrate in the face-down conformation, which is favorable for charge carrier transport. The polymer also showed a good hole mobility and was much higher than that of typical polythiophene structures, such as P3HT. In the last several years, scientists and engineers have heavily explored the properties of PTB polymers. For example, a PTB polymer was used in a BHJ solar cell fabricated from a blend of PTB/PC61BM and showed an efficiency of 4.76% with a V_{oc} of 0.58, a J_{sc} of 12.5 mA/cm^2, and an FF

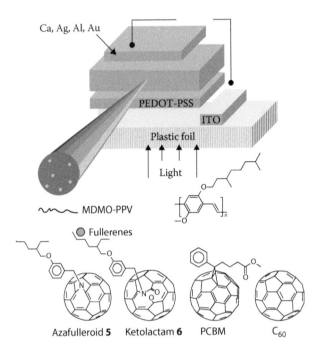

FIGURE 2.8 The molecular structure and device construction of the ketolactam and azafulleroid systems.

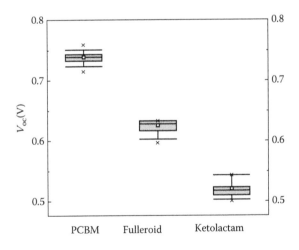

FIGURE 2.9 The open circuit voltages of the fulleroids.

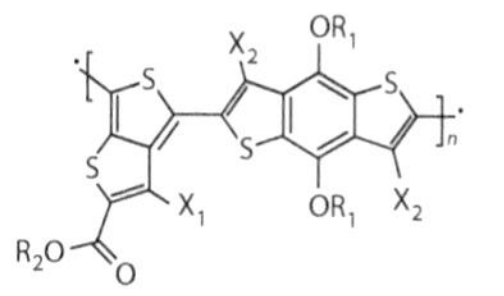

P1 : $X_1 = F$, $X_2 = F$, R_1 = 2-ethylhexyl, R_2 = 2-ethylhexyl
P2 : $X_1 = H$, $X_2 = H$, R_1 = n-octyl, R_2 = 2-ethylhexyl
P6 : $X_1 = F$, $X_2 = H$, R_1 = 2-ethylhexyl, R_2 = 2-ethylhexyl
P7 : $X_1 = H$, $X_2 = F$, R_1 = 2-ethylhexyl, R_2 = 2-ethylhexyl

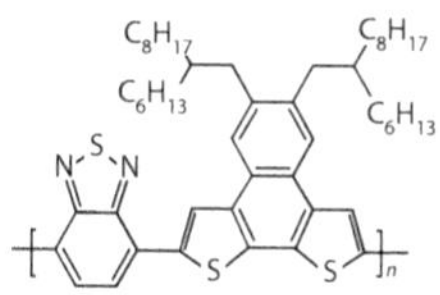

P3 R_1 = 2-ethylhexyl, R_2 = 3-butylnonyl

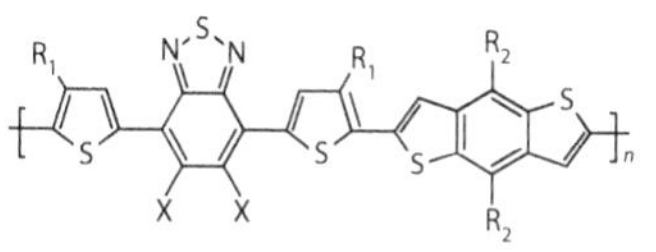

P4 : $X = F$, R_1 = 2-ethylhexyl, R_2 = 3-butylnonyl
P5 : $X = H$, R_1 = 2-ethylhexyl, R_2 = 3-butylnonyl

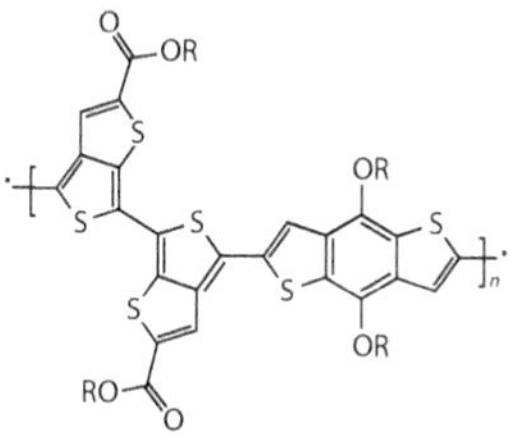

P8 : R = 2-butyloctyl

FIGURE 2.10 The PTB polymers with exceptionally high organic solar cell efficiencies.

of 0.65. The blend of PTB/PC71BM showed a higher efficiency of 5.3%. The properties of PTB were used as figures of merit in the design and creation of new polymers, polymer conditions, and polymer blends (e.g., see Figure 2.10). In one case, one of the blends showed an efficiency of upwards of 6.1%.[79]

This approach with PTB polymers provides an excellent example of structure–function relationships toward improving solar efficiency by fine-tuning the HOMO energy levels of the polymers. For example, in Figure 2.10, for the **P10** (MW of 23.7 kDa) system, the electron-rich alkoxy groups on the BDT moiety were replaced with less electron-rich alkyl chains. The resultant polymer showed a decrease in the HOMO energy level to −5.04 from −4.9 eV with an improved V_{oc} of 0.74 from 0.58 V. Further reduction in the HOMO energy level was achieved by the introduction of fluorine into the TT unit, enhancing its electron-deficiency as shown in the polymer **P11**. The HOMO was observed to decrease with an improvement in V_{oc} to 0.76 V after optimization of device fabrication.[61,70,79,80]

During the last several years, there has been rapid progress in polymer/fullerene BHJ solar cells, which increased their performance beyond 8%.[81] In parallel, there has been clear research progress in understanding and developing donor polymers for achieving high solar cell efficiency. As mentioned earlier, several important guidelines for developing new polymers have been uncovered. There still needs to be a satisfied amount of broad absorption with a high extinction coefficient near the region of maximum solar photon flux, and the polymer still must exhibit a low-lying HOMO energy level and a suitable LUMO energy level. In addition to these concepts, we now know that it is critical that the polymer has appropriate miscibility with *n*-type acceptor materials to form nanoscale bi-continuous interpenetrating networks. The hole mobility of the polymer should be balanced with the electron mobility of the acceptor material for high photocurrent density in solar cell devices. When the polymer satisfies these physical properties, the local dipole moment along the polymer chain is critical for effective exciton separations and charge carrier generation. Apart from the characteristics of the polymer required for high solar cell efficiency, the polymer's photochemical stability and the stability of donor/acceptor nano-morphology are crucial issues to achieve long device lifetimes. It is all of these important characteristics that come into play in the newer highly efficient PTB-type polymers created by Yu and coworkers.

Yu has discovered that some of the best PCE values are obtained with **P6** and **P11** polymers (see Figure 2.10).[18,61,70,79] This might indicate that a suitable π–π distance is important. Good π–π stacking for charge transport has generally been a key criterion for many organic electronic applications involving organic polymers. It is interesting to note that, while fluorine substitution is useful for improving the V_{oc} without sacrificing J_{sc} and *FF*, perfluorination of the polymer backbone (**P1**) can be detrimental to solar cell performance. It is known that perfluorination induces the fluorophobic effect and enhances polymer crystallinity and, therefore, gives a strong driving force for phase separation, leading to the formation of domains on the 100 nm scale. Also, film morphology is affected by the preparation conditions of the active layer for some polymers. Yu et al.[70,79] found that the addition of 1,8-diiodooctane (DIO) to the casting solvent as a processing additive results in a much improved **P6**/PCBM BHJ morphology relative to blended films casted without DIO. Solution phase small-angle x-ray scattering (SAXS) studies have shown that DIO addition to a chlorobenzene solution completely dissolves the PC71BM aggregates, promoting the formation of smaller domains and greater donor−acceptor interpenetration within the film. The iodine atom of DIO has a partial negative charge and PCBM is electron deficient, which may be the reason for their relatively strong interactions with each other and the enhanced solubility of PCBM in the presence of DIO (Figures 2.10 and 2.11). The mechanism by which processing additives control the morphology will be discussed further in the later chapters of this book. Last, as discussed previously, for efficient charge separation and generation, not only the energetics, but also the internal dipole moment along the polymer chain is critical in maintaining the charge transfer characteristics of conjugated polymers.

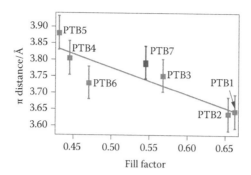

FIGURE 2.11 The correlation of the fill factor with pi-distance in the PTB polymers.

The previously discussed polymers represent some of the best organic conjugated polymer solar cell materials in the field. But there is still so much we can learn about these materials in terms of their structure–function relationships. According to Yu, the current major goal for organic solar cell research is to further enhance efficiencies of cells and modules to above 10% in devices with a large area. The recent rapid development in conjugated polymers seems to indicate that the 10% barrier is within reach. In addition to the development of high-efficiency materials, technologies must also be developed for fabricating cost-effective, lightweight, and flexible devices that show device lifetimes suitable for commercialization. To achieve these goals, interdisciplinary research is required to synergistically address the issues. The developments in this field have led to further scientific investigations of the processes that are operative in organic solar cells. A number of important measurements have made an impact on our understanding of these processes and are the subject of Chapter 3.

REFERENCES

1. A. Leach, Printable, transparent, organic—The future of solar photovoltaics with Germany's KIT, Power-Technology.com, February 2014.
2. C.J. Brabec, *Solar Energy Materials and Solar Cells*, 83, 273–292, 2004.
3. T. Tsutsui and N. Takada, Progress in emission efficiency of organic light-emitting diodes: Basic understanding and its technical application, *Japanese Journal of Applied Physics*, 52, 110001, 2013.
4. Ch. Bosshard, J. Hulliger, M. Florsheimer, and P. Gunter, *Organic Nonlinear Optical Materials*, p. 256, CRC Press, Taylor & Francis Group, Boca Raton, FL. 2001.
5. L. Basabe-desmonts, D.N. Reinhoudt, and M. Crego-calama, Design of fluorescent materials for chemical sensing, *Chemical Society Review*, 993, 2007.
6. Y. Taniyasu, M. Kasu, and T. Makimoto, An aluminium nitride light-emitting diode with a wavelength of 210 nanometres, *Nature*, 441(7091), 325–328, 2006.
7. F. Zhang, D. Wu, Y. Xu, and X. Feng, Thiophene-based conjugated oligomers for organic solar cells, *Journal of Material Chemistry*, 21, 17590, 2011.
8. M. Dolores Perez, C. Borek, S.R. Forrest, and M.E. Thompson, Molecular and morphological influences on the open circuit voltages of organic photovoltaic devices, *Journal of the American Chemical Society*, 131(26), 9281–9286, 2009.

9. X.Z. Yu, K.Y. Wong, and A.F. Garito, Nonlinear optics of organic molecules and polymers, *Introduction to Nonlinear Optics*, p. 1, 1997.

10. A. Yella, H.W. Lee, H.N. Tsao, C.Y. Yi, A.K. Chandiran, M.K. Nazeeruddin, E.W.G. Diau, C.Y. Yeh, S.M. Zakeeruddin, and M. Gratzel, Porphyrin-sensitized solar cells with Cobalt (II/III)-based redox electrolyte exceed 12 percent efficiency, *Science*, 334(6056), 629–634, 2011.

11. P. Peumans, A. Yakimov, and S.R. Forrest, Small molecular weight organic thin-film photodetectors and solar cells, *Journal of Applied Physics*, 93(7), 3693–3723, 2003.

12. Y. Shirota, Organic materials for electronic and optoelectronic devices, *Journal of Materials Chemistry*, 10(1), 1–25, 2000.

13. D. Wohrle and D. Meissner, Organic solar cells, *Advances Materials*, 3(3), 129–138, 1991.

14. S. Gunes, H. Neugebauer, and N.S. Sariciftci, Conjugated polymer based organic solar cells, *Chemical Reviews*, 107(4), 1324–1338, 2007.

15. J.W. Chen and Y. Cao, Development of novel conjugated donor polymers for high-efficiency bulk-heterojunction photovoltaic devices, *Accounts of Chemical Research*, 42(11), 1709, 2009.

16. G. Yu, J. Gao, J.C. Hummelen, F. Wudl, and A.J. Heeger, Polymer photovoltaic cells-enhanced efficiencies via a network of internal donar-acceptor heterojunctions, *Science*, 2705243, 1789–1791, 1995.

17. P.V. Kamat, Meeting the clean energy demand: Nanostructure architectures for solar energy conversion, *Journal of Physical Chemistry C*, 2007, 111(7), 2834.

18. L. Lu, T. Xu, W. Chen, and E.S. Landry, L. Yu, Ternary blend polymer solar cells with enhanced power conversion efficiency, *Nature Photonics*, 8, 716–722, 2014.

19. A. Mishra and P. Buerle, Small molecule organic semiconductors on the move: Promises for future solar energy technology, *Angewandte Chemie International Edition*, 51, 2020–2067, 2012.

20. J. Kalowekam and E. Baker, Estimating the manufacturing cost of purely organic solar cells, *Solar Energy*, 838, 1224, 2009.

21. L.T. Canham, Silicon quantum wire array fabrication by electrochemical and chemical dissolution of wafers, *Applied Physics Letters*, 57(10), 1046, 1990.

22. X.N. Yang, J. Loos, S.C. Veenstra, W.J.H. Verhees, M.M. Wienk, J.M. Kroon, M.A.J. Michels, R.A.J. Janssen, *NanoLetters*, 5(4), 579, 2005.

23. Y.-J. Cheng, S.-H. Yang, and C.-S. Hsu, Synthesis of conjugated polymers for organic solar cell applications, *Chemical Reviews*, 109, 5868–5923, 2009.

24. T.S. van der Poll, J.A. Love, T.-Q. Nguyen, and G.C. Bazan, Non-basic high-performance molecules for solution-processed organic solar cells, *Advanced Materials*, 24(27), 3646, 2012.

25. B. Walker, A.B. Tamayo, X.D. Dang, P. Zalar, H.J. Seo, A. Garcia, M. Tantiwiwat, and T.Q. Nguyen, Nanoscale phase separation and high photovoltaic efficiency in solution-processed, small-molecule bulk heterojunction solar cells, *Advanced Functional Materials*, 19, 1–7, 2009.

26. A. Yassin, P. Leriche, M. Allaina, and J. Roncali, Donor–acceptor–donor (D–A–D) molecules based on isoindigo as active material for organic solar cells, *New Journal of Chemistry*, 37, 502–507, 2013.

27. K. Mahmood, Z.-P. Liu, C. Li, Z. Lu, T. Fang, X. Liu, J. Zhou, T. Lei, J. Pei, and Z. Bo, Novel isoindigo-based conjugated polymers for solar cells and field effect transistors, *Polymer Chemistry*, 4, 3563–3574, 2013.

28. H. Imahori, T. Umeyama, and S. Ito, Large pi-aromatic molecules as potential sensitizers for highly efficient dye-sensitized solar cells, *Accounts of Chemical Research*, 42(11), 1809, 2009.

29. I. McCulloch et al., Liquid-crystalline semiconducting polymers with high charge-carrier mobility, *Nature Materials*, 5, 328–333, 2006.

30. H. Bassler, Charge transport in disordered organic photoconductors, *Physica Status Solidi B—Basic*, 175(1), 15, 1993.
31. P. Peumans, S. Uchida, and S.R. Forrest, Efficient bulk heterojunction photovoltaic cells using small-molecular-weight organic thin films, *Nature*, 425(6954), 158, 2003.
32. G. de la Torre, C.G. Claessens, and T. Torres, Phthalocyanines: Old dyes, new materials. Putting color in nanotechnology, *Chemical Communications*, 2007, 2000, 2007.
33. M. Pfeiffer, K. Leo, X. Zhou, J.S. Huang, M. Hofmann, A. Werner, and J. Blochwitz-Nimoth, Doped organic semiconductors: Physics and application in light emitting diodes, *Organic Electronics*, 4(2), 89, 2003.
34. P.R.L. Malenfant, C.D. Dimitrakopoulos, J.D. Gelorme, L.L. Kosbar, T.O. Graham, A. Curioni, and W. Andreoni, N-type organic thin-film transistor with high field-effect mobility based on a N,N′-dialkyl-3,4,9,10-perylene tetracarboxylic diimide derivative, *Applied Physics Letters*, 80(14), 2571, 2002.
35. B.A. Gregg, Excitonic solar cells, *Journal of Physical Chemistry B*, 107(20), 4688, 2003.
36. C.J. Brabec, A. Cravino, D. Meissner, N.S. Sariciftci, T. Fromherz, M.T. Rispens, L. Sanchez, and J.C. Hummelen, Origin of the open circuit voltage of plastic solar cells, *Advanced Functional Materials*, 11(5), 374, 2001.
37. B.A. Gregg and M.C. Hanna, Comparing organic to inorganic photovoltaic cells: Theory, experiment, and simulation, *Journal of Applied Physics*, 93(6), 3605, 2003.
38. C.W. Chu, Y. Shao, V. Shrotriya, and Y. Yang, Efficient photovoltaic energy conversion in tetracene-C-60 based heterojunctions, *Applied Physics Letters*, 86(24), 243506, 2005.
39. P.W.M. Blom, V.D. Mihailetchi, L.J.A. Koster, and D.E. Markov, Device physics of polymer: Fullerene bulk heterojunction solar cells, *Advanced Materials*, 19(12), 1551, 2007.
40. M. Obarowska, R. Signerski, and J. Godlewski, Generation of charge carrier pairs in tetracene layers, *Macromolecular Symposia*, 212(1), 427–434, 2004.
41. M.-G. Kang, H.J. Park, S.H. Ahn, T. Xu, and L. Jay Guo, Toward low-cost, high-efficiency, and scalable organic solar cells with transparent metal electrode and improved domain morphology, *IEEE Journal of Selected Topics in Quantum Electronics*, 16(10), 1807, 2010.
42. P. Baruch, A. De Vos, P.T. Landsberg, and J.E. Parrott, On some thermodynamic aspects of photovoltaic solar energy conversion, *Solar Energy Materials and Solar Cells*, 36, 201–222, 1995.
43. C.W. Tang, 2-layer organic photovoltaic solar cell, *Applied Physics Letters*, 48(2), 183, 1986.
44. R. Kenneth et al., Re-evaluating the role of sterics and electronic coupling in determining the open-circuit voltage of organic solar cells, *Advanced Materials*, 25, 6076–6082, 2013.
45. G. Witte and C. Woll, Growth of aromatic molecules on solid substrates for applications in organic electronics, *Journal of Materials Chemistry*, 19(7), 1889, 2004.
46. A.S. Paraskar, A.R. Reddy, A. Patra, Y.H. Wjjsboom, O. Gidron, L.J.W. Shimon, G. Leitus, and M. Bendikov, Rubrenes: Planar and twisted, *Chemistry—A European Journal*, 14(34), 10639, 2008.
47. T. Petrenko, O. Krylova, F. Neese, and M. Sokolowski, Optical absorption and emission properties of rubrene: Insight from a combined experimental and theoretical study, *New Journal of Physics*, 2009, 11, 015001.
48. J.G. Xue, S. Uchida, B.P. Rand, and S.R. Forrest, Asymmetric tandem organic photovoltaic cells with hybrid planar-mixed molecular heterojunctions, *Applied Physics Letters*, 85(23), 5757, 2004.
49. M. Dolores Perez, C. Borek, P.I. Djurovich, E.I. Mayo, R.R. Lunt, S.R. Forrest, and M.E. Thompson, Organic photovoltaics using tetraphenylbenzoporphyrin complexes as donor layers, *Advanced Materials*, 21(14–15), 1517–1520, 2009.

50. J.L. Bredas, D. Beljonne, V. Coropceanu, and J. Cornil, Charge-transfer and energy-transfer processes in pi-conjugated oligomers and polymers: A molecular picture, *Chemical Reviews*, 104(11), 4971, 2004.

51. C. Borek et al., Highly efficient, near-infrared electrophosphorescence from a Pt–metalloporphyrin complex, *Angewandte Chemie International Edition*, 46, 1109, 2007.

52. H. Xiang, J. Cheng, X. Ma, X. Zhou, and J.J. Chruma, Near-infrared phosphorescence: Materials and applications, *Chemical Society Reviews*, 42(14), 6128, 2013.

53. T.H. Tran-Thi, J.F. Lipskier, P. Maillard, M. Momenteau, J.M. Lopez-Castillo, and J.P. Jay-Gerin, Effect of the exciton coupling on the optical and photophysical properties of face-to-face porphyrin dimer and trimer: A treatment including the solvent stabilization effect, *Journal of Physical Chemistry*, 96(3), 1073–1082, 1992.

54. G. Gustafsson, Y. Cao, G.M. Treacy, F. Klavetter, N. Colaneri, and A.J. Heeger, Flexible light emitting diodes made from soluble conducting polymers, *Nature*, 357(6378), 477, 1992.

55. J.H. Burroughes, D.D.C. Bradley, A.R. Brown, R.N. Marks, K. Mackay, R.H. Friend, P.L. Burn, and A.B. Holmes, Light emitting diodes based on conjugated polymers, *Nature*, 347(6293), 539, 1990.

56. S.R. Forrest, The path to ubiquitous and low-cost organic electronic appliances on plastic, *Nature*, 428(6986), 911, 2004.

57. R.J. Kline, D.M. DeLongchamp, D.A. Fischer, E.K. Lin, L.J. Richter, M.L. Chabinyc, M.F. Toney, M. Heeney, and I. McCulloch, Critical role of side-chain attachment density on the order and device performance of polythiophenes, *Macromolecules*, 40(22), 7960, 2007.

58. H. Hoppe and N.S. Sariciftci, Morphology of polymer/fullerene bulk heterojunction solar cells, *Journal of Materials Chemistry*, 16(1), 45, 2006.

59. H. Zhou, L. Yang, S. Xiao, S. Liu, and W. You, Donor–acceptor polymers incorporating alkylated dithienylbenzothiadiazole for bulk heterojunction solar cells: Pronounced effect of positioning alkyl chains, *Macromolecules*, 43(2), 811, 2010.

60. S.H. Park, A. Roy, S. Beaupre, S. Cho, N. Coates, J.S. Moon, D. Moses, M. Leclerc, K. Lee, and A.J. Heeger, Bulk heterojunction solar cells with internal quantum efficiency approaching 100%, *Nature Photonics*, 3(5), 297, 2009.

61. H.J. Son, B. Carsten, I.H. Jung, and L. Yu, Overcoming efficiency challenges in organic solar cells: Rational development of conjugated polymers, *Energy and Environmental Science*, 5, 8158, 2012.

62. X. Bulliard et al., Enhanced performance in polymer solar cells by surface energy control, *Advanced Functional Materials*, 20(24), 4381, 2010.

63. C.R. McNeill, S. Westenhoff, C. Groves, R.H. Friend, and N.C. Greenham, Influence of nanoscale phase separation on the charge generation dynamics and photovoltaic performance of conjugated polymer blends: Balancing charge generation and separation, *Journal of Physical Chemistry C*, 111(51), 19153–19160, 2007.

64. S.A. Jenekhe and S. Yi, Efficient photovoltaic cells from semiconducting polymer heterojunctions, *Applied Physics Letters*, 77(17), 2635, 2000.

65. J.Y. Lee, S.T. Connor, Y. Cui, and P. Peumans, Semitransparent organic photovoltaic cells with laminated top electrode, *NanoLetters*, 10(4), 1276, 2010.

66. B.C. Thompson, Y.G. Kim, T.D. McCarley, and J.R. Reynolds, Soluble narrow band gap and blue propylenedioxythiophene-cyanovinylene polymers as multifunctional materials for photovoltaic and electrochromic applications, *Journal of American Chemical Society*, 128(39), 12714, 2006.

67. M.C. Scharber and N.S. Sariciftci, Efficiency of bulk-heterojunction organic solar cells, *Progress in Polymer Science*, 38(12), 1929–1940, 2013.

68. M.M. Alam and S.A. Jenekhe, Efficient solar cells from layered nanostructures of donor and acceptor conjugated polymers, *Chemistry of Materials*, 16(23), 4647–4656, 2004.

69. I.D. Parker, Carrier tunneling and device characteristics in polymer light emitting diodes, *Journal of Applied Physics*, 75(3), 1656, 1994.

70. H.-Y. Chen, J. Hou, S. Zhang, Y. Liang, G. Yang, Y. Yang, L. Yu, Y. Wu, and G. Li, Polymer solar cells with enhanced open-circuit voltage and efficiency, *Nature Photonics*, 6, 649, 2009.

71. J.H. Hou et al., Bandgap and molecular energy level control of conjugated polymer photovoltaic materials based on benzo[1,2-b:4,5-b′]dithiophene, *Macromolecules*, 41, 6012, 2008.

72. J. Hou, H.-Y. Chen, S. Zhang, G. Li, and Y. Yang, Synthesis, characterization and photovoltaic properties of a low bandgap polymer based on silole-containing polythiophenes and benzo[c][1,2,5]thiadiazole, *Journal of American Chemical Society*, 130, 16144, 2008.

73. Y. Huang, L.J. Huo, S.Q. Zhang, X. Guo, C.C. Han, Y.F. Li, and J.H. Hou, Sulfonyl: A new application of electron-withdrawing substituent in highly efficient photovoltaic polymer, *Chemical Communications*, 47(31), 8904, 2011.

74. P. Ravirajan, S.A. Haque, J.R. Durrant, D. Poplavskyy, D.D.C. Bradley, and J. Nelson, Hybrid nanocrystalline TiO_2 solar cells with a fluorene-thiophene copolymer as a sensitizer and hole conductor, *Journal of Applied Physics*, 95(3), 1473, 2004.

75. R.W. Lof, M.A. Vanveenendaal, B. Koopmans, H.T. Jonkman, and G.A. Swatzky, Band gap, excitons, and coulomb interactions in solid C-60, *Physical Review Letters*, 68(26), 3924, 1992.

76. X. Guo, M.J. Zhang, W. Ma, L. Ye, S.Q. Zhang, S.J. Liu, H. Ade, F. Huang, and J.H. Hou, Enhanced photovoltaic performance by modulating surface composition in bulk heterojunction polymer solar cells based on PBDTTT-C-T/PC71BM, *Advanced Materials*, 26(24), 4043, 2014.

77. M.D. Irwin, B. Buchholz, A.W. Hains, R.P.H. Chang, and T.J. Marks, p-Type semiconducting nickel oxide as an efficiency-enhancing anode interfacial layer in polymer bulk-heterojunction solar cells, *Proceedings of the National Academy of Sciences*, 105(8), 2783, 2008.

78. (a) T. Ameri, G. Dennler, C. Lungenschmied, and C.J. Brabeca, Organic tandem solar cells: A review, *Energy and Environmental Science*, 2, 347–363, 2009, (b) C.J. Brabec, A. Cravino, D. Meissner, N.S. Sariciftci, and T. Fromherz, Origin of the open circuit voltage of plastic solar cells, *Advanced Functional Materials*, 11(5), 374, 2001.

79. Y. Liang and L. Yu, A new class of semiconducting polymers for bulk heterojunction solar cells, *Accounts of Chemical Research*, 43(9), 1227–1236, 2010.

80. B. Carsten, J.M. Szarko, H.J. Son, W. Wang, L.Y. Lu, F. He, B.S. Rolczynski, S.J. Lou, L.X. Chen, and L.P. Yu, Examining the effect of the dipole moment on charge separation in donor-acceptor polymers for organic photovoltaic applications, *Journal of American Chemical Society*, 133(50), 20468, 2011.

81. P.P. Khlyabich, B. Burkhart, A.E. Rudenko, and B.C. Thompson, Optimization and simplification of polymer-fullerene solar cells through polymer and active layer design, *Polymer*, 54(20), 5267, 2013.

82. T. Xu and L.P. Yu, How to design low bandgap polymers for highly efficient organic solar cells, *Materials Today*, 17(1), 11, 2014.

83. O.O. Adegoke, I.H. Jung, M. Orr, L.P. Yu, and T. Goodson, Effect of acceptor strength on optical and electronic properties in conjugated polymers for solar applications, *Journal of the American Chemical Society*, 137(17), 5759, 2015.

84. X. Zhang, Z. Li, Z. Zhang, S. Li, C. Liu, W. Guo, L. Shen, S. Wen, S. Qu, and S. Ruan, Efficiency improvement of organic solar cells via introducing combined anode buffer layer to facilitate hole extraction, *J. Phys. Chem. C*, 120, 13954–13962, 2016.

85. O.O. Adegoke, M. Ince, A. Mishra, A. Green, O. Varnavski, M.V. Martínez-Díaz, P. Bäuerle, T. Torres, and T. Goodson III, Synthesis and ultrafast time resolved spectroscopy of peripherally functionalized zinc phthalocyanine bearing oligothienylene-ethynylene subunits, *J. Phys. Chem. C*, 117(40), 20912–20918, 2013.

3 Measurements of Organic Solar Devices and Materials

In order to acquire the details of the function of an organic solar cell, it is important to understand a number of important physical parameters (Figure 3.1). From the particular understanding of how the efficiency of an organic solar cell is obtained to the acquisition of parameters such as absorption, exciton diffusion, and charge transfer character, the experimental success of important techniques has driven the conceptualization and construction of new and improved devices. As this field developed, it was determined that the measurement of the accurate efficiency of an organic solar cell device was not a simple task.[1] Thus, guidelines for the determination of the important physical parameters became a standard. In this chapter, we discuss some of the important experimental procedures used to probe the necessary parameters vital to the function of an organic solar cell device.

STANDARD TEST MEASUREMENTS OF SOLAR CELL EFFICIENCIES

While a large number of new solar materials and cell designs are appearing at a very fast pace, the ability to standardize the quality and magnitude of the efficiency is an important parameter.[1] The measurement of the device efficiency may not be as easy or simple as one might have thought, particularly for multijunction solar cell devices.[2] Over the last 10 years or so, the field has continuingly tried to develop standard protocols for the testing and evaluation of various types of solar cell devices.[2] The U.S. Department of Energy has started to provide such protocols for testing efficiencies, and there are locations around the United States in which one can send particular devices to be evaluated by standardized methods.[3] The reader is highly encouraged to read through the details of documentation that has been given great effort over the years to deliver a good procedure.[3] One such location is the National Renewal Energy Lab (NREL) in Golden, Colorado. NREL holds accreditations in both (1) photovoltaic secondary cell and (2) secondary module and primary reference cell calibration.[4] Using sophisticated instruments and techniques, NREL researchers measure, image, and characterize properties of PV and electronic materials, devices, and interfaces.[4,5] This includes such properties as optical, electrical, material, surface, chemical, and structural performance.[5] Their core competencies include analytical microscopy, device performance, electro-optical characterization, and surface analysis. All of these characterization tools are particularly important in terms of providing accurately calibrated methods to test efficiencies in novel photovoltaic devices. Researchers strive to establish a procedure that can be used as a good model

FIGURE 3.1 **(See color insert.)** Ancient sundial measuring device on a stone platform. (Credit: www.shutterstock.com, 21459691, Tarragona, Spain.)

for testing against other previously reported device efficiencies. The performance of PV cells is commonly described in terms of their efficiency with respect to standard reporting conditions (SRC) defined by temperature, spectral irradiance, and total irradiance.[6] The SRC for rating the performance of terrestrial PV cells are the following: 1000 W/m² irradiance, AM 1.5 (AM: air mass) global reference spectrum, and 25°C cell temperature. The efficiency of a PV cell is given as[7]

$$\eta = \frac{P_{max}}{E_{tot}A} 100 \tag{3.1}$$

where
P_{max} is the measured peak power of the cell
A is the device area
E_{tot} is the total incident irradiance[7]

For Equation 3.1 to give a unique efficiency, E_{tot} must be with respect to a reference spectral irradiance. The current reference spectrum adopted by the international terrestrial photovoltaics community is given in the International Electrotechnical Commission (IEC) Standard 60904-3 and the American Society for Testing and Materials (ASTM) Standard G159.[8,9] A recent improvement to this spectrum is given in ASTM Standard G173.[10] The irradiance incident on the PV cell is typically measured with a reference cell. In practice, for measurements with respect to a reference spectrum, there is a spectral error in the measured short-circuit current (I_{SC}) of the PV cell due to the spectral irradiance of the light source does not match the reference spectrum, which is computer generated, and the spectral responses of the reference detector and test cell are different.[11] This error can be derived based upon the assumption that the photocurrent is the integral of the product of cell responsivity and incident spectral irradiance. Scientists and engineers take great care in calculating this error which can be expressed as a spectral mismatch correction factor (M):[12]

$$M = \frac{\int_{\lambda_1}^{\lambda_2} E_{Ref}(\lambda) S_R(\lambda) d\lambda}{\int_{\lambda_1}^{\lambda_2} E_{Ref}(\lambda) S_T(\lambda) d\lambda} * \frac{\int_{\lambda_1}^{\lambda_2} E_S(\lambda) S_T(\lambda) d\lambda}{\int_{\lambda_1}^{\lambda_2} E_S(\lambda) S_R(\lambda) d\lambda} \quad (3.2)$$

where
  $E_{Ref}(\lambda)$ is the reference spectral irradiance
  $E_S(\lambda)$ is the source spectral irradiance
  $S_R(\lambda)$ is the spectral responsivity of the *reference* cell
  $S_T(\lambda)$ is the spectral responsivity of the test *cell*, each as a function of wavelength (λ)[12]

The limits of integration λ_1 and λ_2 in Equation 3.2 should encompass the range of the reference cell and the test-device spectral responses, and the simulator and reference spectra should encompass λ_1 and λ_2 to avoid error.[12] A matched PV reference cell is typically used as the reference detector, and a solar simulator is used as the light source to minimize the deviation of M from unity. Equation 3.2 is very general and can be applied to most commonly used detectors and light sources. When laser excitation is used in combination with a thermal detector with a wavelength-independent responsivity, the uncertainty in M is dominated by the uncertainty in the spectral responsivity.[12,13] The total effective irradiance of the light source (E_{eff}), which is the total irradiance seen by the cell, can be determined from the short-circuit current of the reference cell under the source spectrum ($I^{R,S}$) from the equation[14]

$$E_{eff} = \frac{I^{R,S} M}{CN} \quad (3.3)$$

where CN is the calibration number for the instrument used to measure the incident irradiance. E_{eff} is different from E_{tot} in Equation 3.1, since E_{tot} usually refers to the total irradiance integrated over the entire spectrum.[14] Both E_{eff} and E_{tot} are derived from integrating $E_S(\lambda)$ over an appropriate range of wavelength. The short-circuit current of a test cell ($I^{T,R}$) at the reference total irradiance (E_{Ref}) is given as[15]

$$I^{T,R} = \frac{I^{T,S} E_{Ref} CN}{I^{R,S} M} \quad (3.4)$$

where $I^{T,S}$ is the short-circuit current of a test cell measured under the source spectrum.[15] Once M is known, the simulator is adjusted so that E_{eff} is equal to E_{Ref}, or[16]

$$I^{T,R} = \frac{I^{R,R} I^{T,S}}{I^{R,S} M} \quad (3.5)$$

TABLE 3.1
Spatial Nonuniformity and Temporal Instability

Class	A	B	C
Spatial nonuniformity	<2	<5	<10
Temporal instability	<2	<5	<10
Total irradiance in 30°	>95	>85	>70

Source: Emery, K., *Solar Cells*, 18(3–4), 251, 1986.
Note: All values are in %.

where $I^{R,R}$ is the calibrated short-circuit current of the reference cell under the reference spectrum and total irradiance.[16] This is the standard simulator-based calibration procedure. The primary terrestrial procedures employed by the United States at NREL follow Equations 3.1 through 3.4 with a primary absolute cavity radiometer as the reference detector and direct normal sunlight as the source spectrum.[17]

The ASTM procedure of the classification of a solar simulator is summarized in Table 3.1.[18] The spatial nonuniformity of a simulator improves as the focal length of the simulator increases.[19] The spatial nonuniformity in and above the test plane is a strong function of the user's ability to properly align the bulb for all arc simulators. The temporal instability of a simulator that mainly affects the current can be improved by correcting real time for intensity fluctuations with an intensity monitor and a computer-controlled data acquisition system. The correction for intensity variations over the I–V measurement period is essential for accurate measurements with a pulsed simulator. Most simulators have 95% of the total irradiance within a 30° field of view.[20] It has been reported that the spectral irradiance for most solar simulators changes with the bulb age, current, and cleanliness of the optics.[21]

The spectral irradiance strongly depends on the optical conditions. The measured direct normal spectral irradiance under natural sunlight and clear sky conditions can be treated as a class B solar simulator for the direct standard spectrum. The same measured direct normal spectral irradiance can be treated as a class A solar simulator for the global standard spectrum.[22] A "typical" clear sky global normal spectral irradiance can be treated as a class A global and direct normal solar simulator. A more general tungsten-halogen bulb with a dichroic filter simulator is a class C solar simulator. This is due to the lack of energy in the 0.4–0.5 μm wavelength interval and too much energy in the 0.7–0.8 μm range.[23] The spectral irradiance of most commercial Xe arc solar simulators are class B or C.[24]

The use of a filtered silicon reference cell can reduce error due to spectral mismatch to less than 1%.[25] Even though tungsten-halogen light sources are poor simulators due to the spectrum shift with age and spatial nonuniformity, the error from spectral mismatch can be reduced to 1% when using a filtered silicon reference

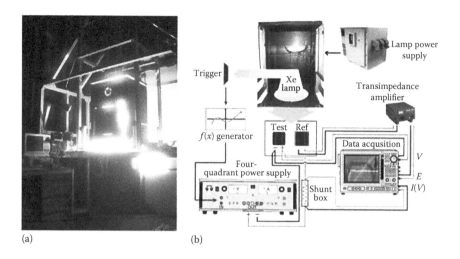

(a) (b)

FIGURE 3.2 **(See color insert.)** Solar simulators used to test OPV devices. (a) Solar simulator used to test PV efficiency. (b) Xe lamp apparatus to measure efficiency.

cell to evaluate an amorphous silicon PV device.[25] New and interesting simulators continue to appear on the market. Some, for example, provide simulated sunlight exposure for steady-state testing of photovoltaic modules.[26] This kind of simulator system can be used for various tests requiring medium or long-term exposure, including photo-induced degradation, power output stabilization, hot-spot testing, and accelerated performance evaluation.[27] They also can be used to meet the requirements of IEC 61215 for crystalline silicon and IEC 61646 for thin-film PV modules (Figure 3.2).[28]

In addition to the combined utilization for solar thermal and photovoltaic modules, another special feature of the modern test bench is an automatic shade facility.[28] This facility allows a scientist to expose the photovoltaic modules to radiation for a specific period of time and can determine the time constants of thermal collectors (Figure 3.3).

Once the solar simulator intensity has been set to match the standard intensity and spectrum using a reference cell for the particular device being evaluated, the *I–V* characteristics can be measured. If the voltage sweep rate for the PV device being evaluated is too large, then the fill factor and efficiency can be artificially high.[29] The cell is typically mounted on a temperature-controlled plate with the junction temperature at the standard test temperature (25°C).[30] The plate, temperature can be controlled by gas, water, or most reliably with a thermoelectric module.[30] Typically, one makes contact with the PV device with a vacuum plate, which can be achieved by mounting a plated Ni- or Au-printed circuit board (voltage contact) in a slot in the vacuum plate (current contact), by mounting a spring-loaded probe (voltage) in the vacuum plate (current), or by using Kapton plated with patterned Cu or Ni for electrically isolated voltage and current contacts to the device substrate.[18,25,31] A variety of probe designs available for making contact to PV devices are now readily available.[32]

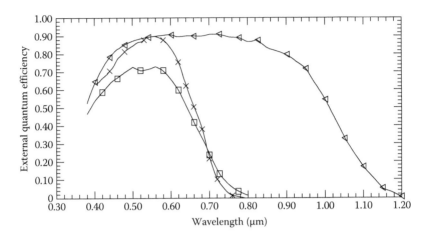

FIGURE 3.3 External quantum efficiencies for a single crystal silicon reference cell (triangle), filtered silicon reference cell (rectangle), and a typical amorphous silicon test device (cross). (From Emery, K., *Solar Cells*, 18(3–4), 251, 1986; Osborne, M., Konica Minolta's AK Series photovoltaic reference cells provide consistent cell measurements, PV Tech, http://www.pv-tech.org/product_reviews/konica_minoltas_ak_series_photovoltaic_reference_cells_provide_consistent_c, London U.K., 2011.)

The fill factor and efficiency can be artificially enhanced by separating the voltage and current contacts or by adding multiple current contacts making poor grid designs appear optimal. Too small a current contact area can cause current crowding (localized heating) reducing the fill factor and efficiency, an effect observed in many organic and inorganic electronic devices.[33] Whenever possible, the resistance between the voltage and current contacts should be measured when making contact to the device. Custom test fixtures are useful for the rapid evaluation of PV devices in a production environment. Silver paste or silver epoxy is sometimes used because it provides a convenient means of attaching wires for subsequent testing. Attaching wires with solder or an ultrasonic bonder is essential for evaluating concentrator cells because of the large currents.[34] The use of metallized rubber is popular in evaluating some cells because large area contacts can be made to thin metallizations without damage.

The reader should appreciate the fine details in the art of evaluating the efficiency of a solar cell. Indeed, great care must be taken not only in the setup of the apparatus for the evaluation but also for the materials used. The analysis of the spectral response and current dependence requires great experience in determining the differences in responses from one cell to another. In some cases, this can be relatively complicated by the absorption and emission processes of the materials used. As we will see in later chapters, there have been disagreements in the literature of particular efficiencies of specific materials due to varying test measurement approaches. The field is still moving toward an agreed unified procedure (with similar test equipment) for the evaluation of all organic solar devices to alleviate this problem.

KELVIN PROBE SPECTROSCOPY OF ORGANIC SOLAR CELL MATERIALS

The reader should now appreciate the sensitivity in the device measurement and the effort required to obtain reliable parameters. Most researchers in this area believe the materials characterization is the initial crucial step in evaluating a potential device's operation. One parameter that is key in this regard is the measurement of the potential across the device. This has led to measurements of the uniformity of the potential across the surface of the film in an organic solar cell. One such method to carry out this measurement is Kelvin probe force microscopy (KFM), which is based on atomic force microscopy (AFM). It was proposed by Nonnenmacher in 1991, and allows us to obtain not only topographic images but also potential images.[34,35] The term Kelvin is used for the reason that the principle of KFM is analogous to that of the Kelvin method. A potential is obtained by detecting a cantilever deflection caused by an electrostatic force between a tip and a sample. There is a contact potential difference (CPD) between a tip and a sample, namely, the work function difference between a tip and a sample, as shown in Figure 3.4. Here, an AC modulation bias V_{AC} with a DC offset bias V_{DC} between the tip and the sample to generate an electrostatic force between the tip and the sample (see Figure 3.5).[35] The cantilever deflection by an electrostatic force is detected by a photodetector, and then the signal of frequency f_{AC} is collected by a lock-in amplifier. The signal is transferred to a feedback controller, and the CPD is obtained by adjusting the dc offset bias V_{DC} so that the component signal of frequency f_{AC} becomes zero.[36,37]

One can understand this process by considering the bandgaps of the sample and tip as shown in Figure 3.5. The bandgap between the vacuum and the Fermi levels, namely, the work function, differs from one material to another. It is by this convention that the work functions of the tip and the sample are defined as ϕ_1, ϕ_2, respectively as shown in Figure 3.5. The electric field $\phi_2 - \phi_1$ is generated between the tip and the sample. By applying an ac modulation bias V_{AC}, the induced electrostatic force F_{ES} between the tip and the sample by the electric field is given by[38,39]

$$F_{ES} = \frac{1}{2} * \frac{dC}{dz} \left\{ V_{DC} + V_{AC} \sin\left(2\pi f_{AC}t\right) - \left(\phi_2 - \phi_1\right) \right\}^2 \qquad (3.6)$$

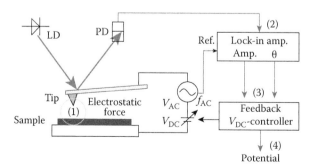

FIGURE 3.4 The Kelvin force feedback loop.

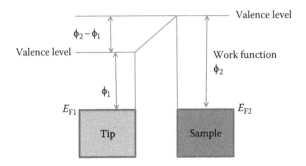

FIGURE 3.5   The band diagram of the tip and sample in KFM method.

where C and z are the capacitance and distance between the tip and the sample, respectively, and this equation leads to[39]

$$F_{ES} = \frac{1}{2} * \frac{dC}{dz} \left[\frac{\left\{ V_{DC} - \left(\phi_2 - \phi_1 \right)^2 \right\}}{2} + \frac{V_{AC}^2}{4} \right]$$

$$+ \frac{1}{2} * \frac{dC}{dz} * V_{AC} \left[V_{DC} - \left(\phi_2 - \phi_1 \right) \right] \sin \left(2\pi f_2 t \right) - \frac{1}{2} * \frac{dC}{dz} * \frac{V_{AC}^2}{4} \cos \left(4\pi f_2 t \right) \quad (3.7)$$

If one applies a dc offset bias V_{DC} between the tip and the sample, which is equal to the electric field $\phi_2 - \phi_1$ between the tip and the sample, the vacuum levels of the tip and the sample arrive at the same height, and simultaneously the second term of the right side of the equation 3.7, namely, the f_2 component signal, becomes zero. Therefore, we can obtain the intended potential by adjusting the dc offset bias V_{DC} to nullify the f_2 component signal.[39]

Similar to this description, the Kelvin probe force microscopy (KFM) technique adopts a noncontact tip with a conductive coating to measure the potential difference between the tip and the adjacent surface.[39] In general, the surface potential relates to a number of different surface phenomena. This can include catalytic activity, doping/impurities/defects, band bending at interfaces, and the polarization of the surfaces.[40,41] The assessed surface potential with a resolution of a few nanometers produced by KFM gives useful information about the composition, surface charge distribution, and electronic states on the surface of a solid. A great number of recent studies have utilized KFM to obtain both structural and electronic properties of conjugated polymer-based photovoltaic materials.[43,44] The ability of the KFM method to obtain quantitative mappings in a noncontact and nondestructive manner makes it very attractive for the analysis of organic materials.[45] The technique was also later to be found useful for the purposes of measuring the work function of thin heterostructures, thus enabling its use for catalytic applications.[46] Furthermore, correlation between shifts in the surface potential (surface potential difference in the presence or absence of illumination) and the power conversion efficiency of polymer photovoltaic devices based on poly(3-hexyl-thiophene) (P3HT) and titanium dioxide nanorod hybrid bulk heterojunctions

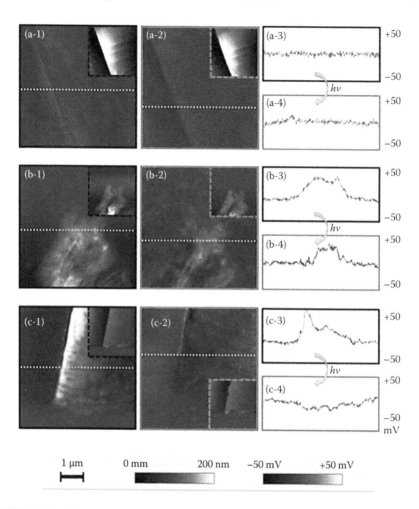

FIGURE 3.6 (See color insert.) The Kelvin probe measurement of the polymer. Surface potential mappings in the dark (a-1, b-1, c-1) and under UV-B illumination (a-2, b-2, c-2) of three kinds of nano materials (a) SHTNTs, (b) N-TiO$_2$ NWs, and (c) N-TiO$_2$-P+NWs on gold thin film. The insets of surface potential images are topographic images (a-3, b-3, c-3) and (a-4, b-4, c-4) are surface potential values of the cross section obtained from the corresponding white-dotted line without and under UV-B illumination respectively. (From Wu, M.-C. et al., *J. Nanopart. Res.*, 16, 2143, 2014.)

could also be assessed by KFM.[42,47,48] When applied in ultrahigh vacuum, an improvement of the spatial resolution by a factor of 10 has been reported (Figure 3.6).[39]

IMPEDANCE SPECTROSCOPIC MEASUREMENTS IN ORGANIC SOLAR CELLS

While AFM and KFM are excellent techniques to probe the topography and parameters of a solar cell device, they are mainly static measurements. When providing information to the scientists regarding the operation of a particular device, one usually considers a model. It is commonly accepted that modeling an electrical system mainly consists of establishing a set of structural relationships between suitable parameters which represent its electrical response.[49] But in many cases, the role of instrumentation in the process of constructing the model is not explicitly given consideration. Measurements are always a part of the model-creating process as modeling and measuring become parallel activities. In the specific case of photovoltaic devices, the sole current–voltage response contains poor information about the kinetic mechanisms limiting the solar cell performance.[50] Impedance spectroscopy goes beyond static measurements by applying a small oscillating perturbation to the given steady state of the device operation.[51] Due to the fact that the frequency of the oscillation is changed during the experiment, separate structural parts of the device are able to electrically respond. This is why impedance measurements can be regarded as spectroscopy. The important parameter that gives us information about the number and dynamics of charge carriers in the cell is the capacitance. This is because in all solar cells, the generation of positive and negative carriers creates a splitting of Fermi levels that is ultimately responsible for the photovoltage.[52] For example, in silicon solar cells, the capacitance shows two main components as a function of the bias. The first is a Mott–Schottky characteristic which is due to the modulation of the Schottky barrier at reverse and moderate forward bias. The second component is the chemical capacitance that increases exponentially for intense forward bias.[53–55] The first characteristic indicates the presence of doping, whereby the solar cell device is able to accumulate substantial minorities. Such carrier storage is manifested in the second characteristic, the chemical capacitance, which directly reflects the carrier statistics.[56] Scientists and engineers use these characteristics both qualitatively and quantitatively in order to make predictions about charge carrier effects in organic solar cells.

In the initial constructions of organic dye-sensitized solar cells that were based on nanostructured TiO_2, the shape of the voltage dependence of the capacitance is somewhat different than in the case for silicon. This is due to the fact that the hole conductor is a liquid electrolyte with high ionic concentration, so that the Schottky barrier in the active semiconductor layer is not found.[57] However, the chemical capacitance is very clearly observed and shows the density of states of electrons accumulated in the electron-transporting material (TiO_2).[58] The identification of the voltage-dependent capacitance has become, then, a major tool for assessing the energetics in an organic solar cell and for interpreting the recombination lifetime.[59] The strong accumulation associated with the unconstrained rise of the Fermi level with bias appears in high-performance solar cells, while in many other cases, the charge storage is inhibited by additional mechanisms and the solar cell capacitance may become negative at strong forward bias.

Charge carrier diffusion and recombination in an absorber blend of poly(3-hexylthiophene) (P3HT) and [6,6]-phenyl C61-butyric acid methyl ester (PCBM) with indium tin oxide (ITO) and aluminum contacts have been analyzed in the dark by means of impedance spectroscopy (shown in Figure 3.7).[51] As Figure 3.7 shows, the measured electrical impedance is illustrated in the complex plane for different forward bias voltages. This type of impedance pattern belongs to the

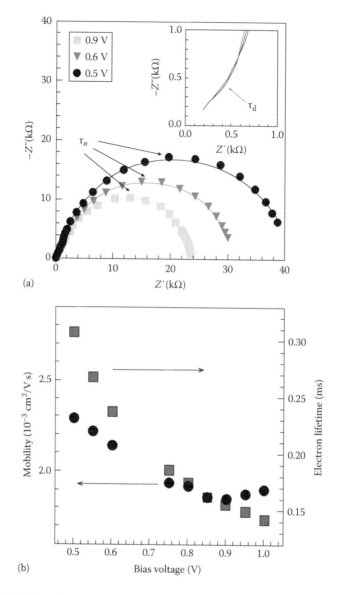

FIGURE 3.7 Impedance spectroscopy of P3HT polymer blend solar cell. (From Garcia-Belmonte, G. et al., *Organ. Electron.*, 9(5), 847, 2008.)

responses usually encountered in systems in which carrier transport is determined by diffusion–recombination between nonabsorbing contacts.[60] Researchers found that for this device system, the reverse bias capacitance exhibits Mott–Schottky-like behavior indicating the formation of a Schottky junction (band bending) at the P3H:PCBM-Al contact.[59] Impedance measurements show that minority carrier (electrons) diffuse out of the P3HT:PCBM-Al depletion zone, and their accumulation contributes to the capacitive response at forward bias.[59] Thus, a diffusion–recombination impedance model accounting for the mobility and lifetime parameters was discussed to explain this effect. Models such as this have been critical in explaining the details of the operation of organic solar cells. Also, it has been shown that impedance spectroscopy can also give important information about the lifetime of free electrons in the device. In general, the lifetime is composed of a trapping factor and a free electron lifetime.[60] Since the diffusion coefficient contains the reciprocal of the trapping factor, it is found that its product (diffusion coefficient) × (lifetime) reveals the shape of the free electron lifetime, which contains the essential information of kinetics of electron transfer at the surface. The use of various models based on an exponential distribution of surface states provides a good description of the voltage and temperature dependence of free electron lifetimes and diffusion lengths in organic solar cells. The impedance measurement is increasingly being utilized in this field in order to probe the electron dynamics in real devices. This is in contrast to the measurements of estimates of these dynamics in solution. While both are extremely important, one would like to have the ability to construct a device and probe the dynamics directly in the solid state.

MEASUREMENT OF EXCITON DIFFUSION IN ORGANIC SOLAR CELLS

While the measurement of all of the important parameters of an organic solar cell are critical, there are many researchers in the field that heavily focused on the exciton diffusion length of the device.[61] The exciton diffusion length L_D in organic polymeric films is often measured using the diffusion-limited quenching at the interface with either fullerenes or fullerene derivatives or metals. The dependence of the relative quenching efficiency, Q, on the polymer thickness, L, is recorded experimentally, and it is fitted with a mathematical model based on a diffusion equation. Such a fitting results in a value of L_D typically in the range of 5–7 nm for many conjugated polymers (e.g., PPV derivatives).[62–67] Researchers have measured the exciton diffusion length to estimate the efficiency of the exciton quenching in the outermost layer of the device structure. The exciton diffusion length is an intrinsic property of the organic material and should be the same in the heterostructures and in pristine films.[65] Since, in a typical sample geometry, the effect of interface quenching enters the diffusion model as boundary conditions, in practice one usually will estimate the efficiencies of such quenching simply by choosing the boundary conditions that lead to the proper value of L_D for both types of samples. The exciton density, n, is modeled by the following diffusion equation:[64–67]

$$\frac{dn(x,t)}{dt} = D * \frac{d^2 n}{dx^2} - \frac{n(x,t)}{\tau} + G(x,t) - S(x)n(x,t) \tag{3.8}$$

where
  τ denotes the exciton lifetime
  D is the exciton diffusion coefficient that is related to the diffusion length by
    the equation $L_D = \tau D$

Due to the sample symmetry, n depends only on one spatial coordinate, which is the distance from the free interface. Because ultrafast photo excitation has often been used, the generation term $G(x,t)$ can be represented as the initial exciton distribution, which is taken to be uniform due to the low absorption coefficient at the excitation wavelength.[62–69] The term $S(x)$ in Equation 3.8 denotes the interface quenching; it can be adjusted to mimic different boundary conditions. The polymer–vacuum interfaces of the heterostructures and the pristine films are assumed to be equal, implying common boundary conditions at this interface. As we will see in later chapters, the characterization of this interface is a critical issue in evaluating the exciton diffusion lengths.

The polymer–fullerene interface in many reported organic solar cells is known to be an efficient exciton quencher, which provides the second boundary condition for the heterostructures with sufficient polymer thickness. The polymer–substrate interface is expected to be a weak exciton quencher; therefore, it is assumed that its quenching efficiency is negligible. With these boundary conditions, Equation 3.8 can be simplified and solved analytically for both types of samples. Then, the relative quenching efficiency Q is simply[65,68]

$$Q(L, L_D) = 1 - \frac{\int_0^L dx \int_0^\infty n_{\text{quenched sample}}(x,t)\, dt}{LN_0\tau} \qquad (3.9)$$

where $LN_0\tau$ is the total photoluminescence of a quencher-free sample. Integration of Equation 3.9 leads to the analytical expressions for the quenching efficiencies for the heterostructures and for pristine films. Scientists in this area have used this methodology to investigate the exciton diffusion lengths in important organic material devices.[71–81] In particular, transient photoluminescence measurements carried out on such well-known conjugated polymers as poly(p-phenylenevinylene) have observed the important process of exciton diffusion to quenching centers by analyzing the nonexponential decay dynamics.[65] It has been shown that the photoluminescence decay rate is essentially determined by the defect concentration in even slightly oxidized samples. The photo-oxidation, however, is much more severe at the polymer film's surface where the excitation light is absorbed.[65] As a consequence, the observed photoluminescence decay is longer and the photoluminescence quantum yield is higher for excitation polarized perpendicular to the chains in oriented samples where the absorption depth is longer.[70,71] This penetration depth effect also explains how the emission lifetime observed in oxidized samples can be longer for relatively penetrating excitation wavelengths than more strongly absorbed colors.[72] In many cases, the determination of the exciton diffusion length can be complicated by the existence of other limiting processes in the device. As observed in Figure 3.8, researchers have found that the interface and substrate effects can alter the estimation of the exciton diffusion.[62] The combination of Forster energy transfer with exciton diffusion in slightly thicker films can lead to a false estimation of the exciton diffusion length.[62] In addition to this, the photodegradation of a material

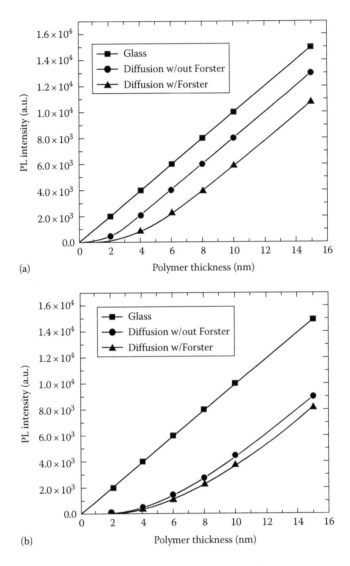

FIGURE 3.8 The PL exciton diffusion quenching experiment for an organic polymer. (From Scully, S.R. and McGehee, M.D., *J. Appl. Phys.*, 100(3), 034907, 2006.)

can also have a profound effect on the exciton diffusion length measurement.[73–78] In other cases, this process can be focused on particular types of chromophores or polymers or even side groups. It was found, for example, that the presence of these C=O groups is inversely correlated to the photoluminescence quantum yield for PPV-type conjugated polymers even in samples that have not been deliberately photo-oxidized. The measurement of exciton diffusion continues to be a critical factor that most researchers believe limit the full introduction of organic materials into the solar cell commercial market.[79–90] As the reader will find, there is a large amount of references devoted to the problem of exciton diffusion in organic solar cells, and we will discuss more about this important process in the interface chapter.

FEMTOSECOND TRANSIENT ABSORPTION SPECTROSCOPY OF ORGANIC SOLAR CELL MATERIALS

The kinetics of the processes that effect the exciton diffusion process are on relatively fast time scales. Time-resolved absorption measurements have also been used to probe the dynamics of excitations in organic PV materials.[91–95] Pump-probe ultrafast techniques have been reported extensively in the literature and have been carefully adapted to performing measurements with the organic PV molecular systems. Some provide the entire spectral time response or for more detailed measurements of time resolution at a specific spectral window, fast pulses at specified wavelengths have been utilized. For example, in the 400 nm region, the use of ultrafast pulses less than 50 fs has been invoked to probe the dynamics of electron and charge transport in thiophene systems. As discussed in Chapter 2, the use of thiophene polymers was critical to the initial discoveries of the solar efficiencies of organic polymers. Understanding the important electronic processes at various time scales is important in predicting better materials with higher solar efficiencies.[91–101] The use of ultrafast lasers has dramatically changed the ability to probe important fast processes in organic solar materials. Most of the lasers have a fundamental wavelength in the near-IR spectral region (near 800 nm). However, in many cases, second harmonic light (from the oscillator or through a cavity-dumped) from a Ti–sapphire oscillator is used. Amplified systems can be utilized in combination with an optical parametric amplifier to gain complete visible and infrared spectral responses. The details of such laser and detection equipment are outside the focus of this book. However, these details can be found in the large number or reports on time-resolved optical properties of organic solar cell materials discussed later. In almost all cases, an optical delay is again used to translate the probe beam and the change in $\%T$ is measured by a fast photodiode. The time resolution of the experiment is characterized by the instrument response function. While many systems vary, most OPA systems use an amplified fs pulse system which contains 1 mJ, 100 fs pulses at 800 nm with a repetition rate of 1 kHz. The output of the OPA is often split to generate pump and probe beam pulses with a beam splitter. For experiments with many organic materials, the pump beams are focused onto a quartz cuvette containing the sample. The probe beam is delayed with a computer-controlled motion controller and then focused into a sapphire plate to generate a white light continuum. The white light is then overlapped with the pump beam in the sample cuvette, and the change in absorbance for the signal is collected by a charge-coupled device (CCD) detector. Data acquisition is controlled by specialized software. The typical power of the probe beam is <0.1 nJ and that of the pump beam is around ~0.1–0.4 nJ/pulse. The sample is stirred with a rotating magnetic stirrer. It is usually important to check for photodegradation of the sample at this point as many organic PV materials under these conditions may undergo photochemical processes that are unwanted (Figure 3.9).

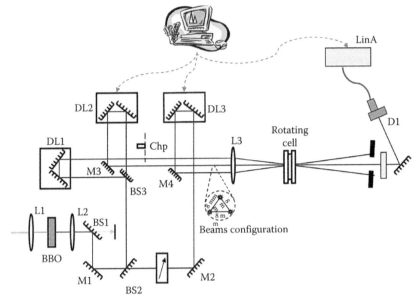

$\lambda_{pump} = \lambda_{probe} = 415$ nm, pump pulse energy ~0.5 nJ; $\Delta A/A \sim 10^{-7}$ S/N > 5

FIGURE 3.9 (**See color insert.**) The optical apparatus for femtosecond transient absorption. (From Varnavski, O.P. et al., *J. Am. Chem. Soc.*, 124, 1736, 2002.)

There are several specific aspects of materials and device physics that can be obtained from the transient absorption measurements. In general, this and other ultrafast absorption techniques can be used for investigations of energy transfer and electronic coupling dynamics in organic polymeric or small molecular solar cell materials (Figure 3.10). For example, detailed transient absorption anisotropy decay measurements have been reported to capture the important dynamics associated with the fast interchange between energy levels that has been observed in selected organic PV systems in the past.[103] It is important to probe molecular systems in a systematic fashion as the properties will vary with very minor changes in the molecular structure of the PV material. For this reason, some scientists investigate the transient absorption properties of organic systems in an iterative manner. For example, in one report both the monomer and trimer systems' transients were obtained to probe the molecular response of a larger macromolecular system.[103] Not only does the use of transient absorption decays provide information regarding the lifetimes of excited states and the effect of charge transfer, but the use of transient absorption anisotropy decay measurements can also probe the mode of energy transfer in the molecular system. By evaluating the decay profile of the anisotropy, one can probe the mode of energy transfer. For example, in one trimer system the reported the transient absorption anisotropy signal decays to a residual value within the first ~50 fs of the excitation.[103,104] It was suggested that within the first part of the excitation pathway the energy is delocalized over the three arms of the trimer system.[103]

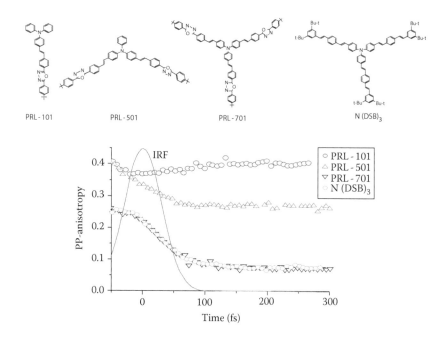

PRL - 101   PRL - 501   PRL - 701   N (DSB)$_3$

FIGURE 3.10 The ultrafast transient absorption of PV materials. (From Varnavski, O. et al., *J. Chem. Phys.*, 116, 8893, 2002.)

It should be noted that with the transient absorption measurement, the FWHM of the instrument response function is significantly shorter than that found in other ultrafast measurements. Very well characterized deconvolution procedures allow for the evaluation of very fast dynamics and a considerably higher degree of accuracy in regards to the initial value of the anisotropy signal.[105] Reports have illustrated the importance of excited state absorption in connection with the initial value and decay of the anisotropy.[106] However, the fast decaying anisotropy is predominantly associated with the fast energy migration process of the three chromophores in the trimer for example in Figure 3.10.[103] The time-resolved results suggest that while at longer time scales the monomer and trimer possess similar dynamics, at shorter time scales, the branching structures demonstrate a fast energy redistribution process that is not observed for the monomer system.[102] This information is useful in designing new macromolecules for the efficient harvesting and transfer of solar energy. The dynamics of the process give further details regarding the mechanism of energy transfer of the important parts of the organic solar cell. The connection of this fast energy redistribution process to the enhancement of organic solar cell efficiency may relate to delocalized states before the final stabilization of the charge transfer process. The connection of the dephasing (and coupling to nuclear motion) to the delocalized state is important in further understanding the distinguishing characteristics of the macromolecular system in comparison to it's smaller molecular parts. There are a number of important energy relaxation processes that will contribute to the anisotropy decay, and these processes are critical to the dynamics of excitons operating in organic PV cells.

FLUORESCENCE UPCONVERSION SPECTROSCOPY AND ANISOTROPY MEASUREMENTS IN ORGANIC SOLAR MATERIALS

As mentioned in the introduction, in order to understand some of the fundamental properties of the solar cell device, it is first necessary to probe the basic properties of the active material. One such property is the fluorescence observed from the material and its lifetime. Measurements of the lifetime of the material may be obtained from such methods as time-correlated single photon counting as well as with the use of a streak camera.[107] These measurements typically produce good time resolution in the nanosecond and several picoseconds time regions. However, for the case of organic systems, this time resolution might not be as accurate. Faster time responses may be obtained from optical gating techniques such as fluorescence upconversion.[108] For example, a number of reports have demonstrated that the time-resolved fluorescence of the organic PV systems may be studied by femtosecond upconversion spectroscopy.[109] For studying organic materials, the upconversion system used in most experiments invokes both the second harmonic and the third harmonic generation of an fs oscillator to generate (1) 420 nm, (2) 377 nm, and (3) 266 nm, respectively, separately from a mode-locked Ti–sapphire laser. The gate step size can be as fast as 6.25 fs. Spectral resolution is achieved by dispersing the upconverted light in a monochromator and detecting it by the use of a photomultiplier tube (Figure 3.11).

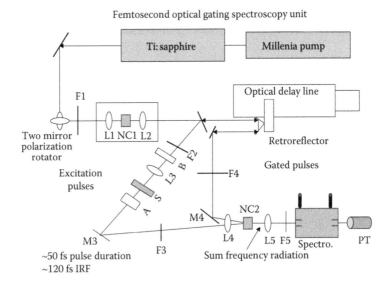

FIGURE 3.11 Measurement apparatus for fluorescence decay tests of PV materials. (From Ransinghe, M. et al., *J. Phys. Chem.*, 108, 8543, 2004.)

The average excitation power is kept at a low level in order to not oversaturate the excitation or photodamage the organic material.[109] At these excitation intensities, the fluorescence dynamics is usually found to be independent of the excitation intensity, for all investigated solutions. The sample is excited with frequency doubled or tripled light from a mode-locked Ti–sapphire laser. The luminescence emitted from the sample is upconverted in a nonlinear crystal of β-barium borate using the pump beam at about 800 nm that is first passed through a variable delay line. For visible excitation (385–415 nm with a typical fs oscillator), the fundamental (770–830 nm, 40 fs, 82 MHz) pulse from the oscillator is frequency doubled. The sum frequency signal is collected as a function of the delay. This maps the fluorescence decay of the PV material. What is also often measured is the system's capability to carry out ultrafast emission anisotropy decay measurements with very high time resolution.[110] In this case, the polarization of the excitation beam for the anisotropy measurements is controlled using a Berek compensator, and the rotating sample cuvette was 1 mm thick. Lifetimes are obtained by convoluting the signal with the instrument response function (Figure 3.12).

The spectral resolution is achieved by using a monochromator and PMT. Anisotropy measurements may be obtained from repeating the fluorescence lifetime measurements with parallel excitation and parallel emission ($I_{\parallel}$) and again with perpendicular excitation and parallel emission ($I_{\perp}$). The total (rotation-free) intensity

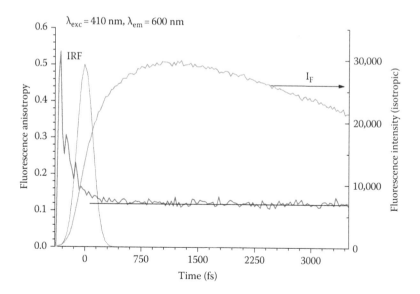

FIGURE 3.12 The fluorescence decay profile of an organic PV material using femtosecond fluorescence upconversion. (From Goodson, T., *Annu. Rev. Phys. Chem.*, 56, 581, 2005.)

decay is given by $I(t) = I_{\parallel}(t) + 2I_{\perp}(t)$.[112] Anisotropies as a function of time are then presented as a result of the following relationship:[112]

$$r(t) = \frac{I_{\parallel}(t) - I_{\perp}(t)}{I_{\parallel}(t) + 2I_{\perp}(t)} \tag{3.10}$$

In many cases, a standard (e.g., coumarin 30) is used to provide a G-factor which is needed for the analysis of these measurements. Data fitting and deconvolution at less than 10 ps can be accomplished with rather high accuracy using common software packages. For example, the maximum deconvolution time resolution may give a minimum resolution of roughly one-sixth of the cross-correlation function, or 30 fs.[109] Data fitting at longer times may be performed using normal graphical software. Generally, for these organic PV materials the emission lifetimes are the result of multiexponential decay fitting. When contributions to total emission are presented for a given relaxation lifetime as a percentage, the percentage is given as a ratio of the separated and deconvoluted integrations for the decay, weighted by the fitted amplitude for each decay, and compared to total emission.

With novel conjugated small molecules, polymers, and dendrimer systems, the process of energy migration using ultrafast fluorescence anisotropy decay measurements has been investigated.[109–113] In many cases, these materials were candidates for organic PV as well as LED materials.[114] Fluorescence anisotropy decay measurements for a number of organic conjugated systems provided new insight into the strength of intramolecular interactions in particular organic systems and their connection to the solar cell's efficiency. For example, with the good time resolution of the upconversion technique, it is possible for researchers to capture the dynamics within the first picosecond (<1 ps) of relaxation, presumably during the redistribution of energy. Certain organic macromolecules showed an initial fast anisotropy decay component of ~70 fs (with convolution fitting procedures). Longer time components are also detected but are attributed to the rotational motion of the entire branched structure and not due to fast energy migration processes that are critical to the description of intramolecular interactions (Figure 3.13).

It appeared that this fast delocalization of energy mainly concerned the excitation of chromophores in a given macromolecule is a redistribution of this energy to the surroundings. To understand the importance of the electronic structure, geometry, and polarizability of organic solar cell materials, researchers have carried out measurements with a number of different geometrical structures with similar building block chromophores which are useful components for organic solar cells.[115–120] For example, in the case of branched macromolecules where the chromophores are ligands around a branching center, one can alter the branching center with nitrogen (N), carbon (C), phosphorous (P), or benzene (meta substituted). This system illustrates the power of this technique and the process of energy migration by ultrafast fluorescence anisotropy measurements.[121] These results gave specific information regarding the relative strength of interactions in the particular branching center. It was found that the nitrogen-centered system provides the strongest intramolecular interaction among chromophores by virtue of the highest initial depolarization rate. While the initial anisotropy value was obscured by the finite width of the instrument

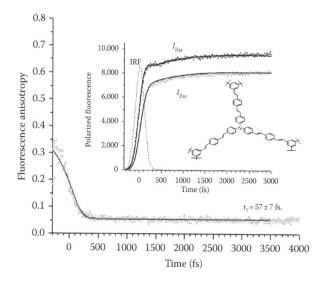

FIGURE 3.13 (**See color insert.**) The fluorescence anisotropy of organic dendrimers for PV. (From Ransinghe, M. et al., *J. Phys. Chem.*, 108, 8543, 2004.)

response function, fitting procedures suggested that the first initial decay process was complete within the first 100 fs of the decay profile.[114] The residual value close to ~0.05–0.1 was suggested as the point of the onset of the slower rotational motion component of the depolarization decay process. With similar chromophores attached, other branched centered systems showed significantly different anisotropy dynamics compared to the nitrogen-centered system. It became clear that the relative interaction as expressed by fast depolarization measurements can be altered synthetically by placing the appropriate building blocks in a particular orientation (in principle) as well as the appropriate use of branching centers to tune the interaction among chromophores.[121]

Researchers have used these ultra-fast techniques to look into the dynamics of larger solar cell macromolecular material as well. Investigations with three generations of the triphenylamine dendrimer suggested that it is possible in these systems that the length of the delocalized excitation indeed could be increased without the use of fully linear conjugated linkages between branches.[122] This information is critical to larger organic PV-covered portions of the solar spectrum. Time-resolved fluorescence anisotropy measurements have also demonstrated these and other interesting features of strong intramolecular coupling. The anisotropy decay had two different features that describe the relative strength of the intramolecular interactions and the effect of increasing the number of participating chromophores. What was found was that there was an increasing amplitude of a fast (~50 fs) time component in the decay profiles with an increase in dendrimer generation. The coherent excitation mechanism may become dominant for the case of larger dendrimer generations. Also, the residual value of the anisotropy decay (the decay to the point at which rotational diffusion would persist) decreased for increasing dendrimer generation. As the number of participating chromophores increased, the final depolarized emission component resulted in a smaller residual anisotropy value. This and other related ultra-fast optical studies strongly suggest that it is possible to build larger macromolecular systems, which may still maintain the effective electronic coupling over a number of chromophores.

TWO-PHOTON EXCITED FLUORESCENCE MEASUREMENTS IN ORGANIC SOLAR CELL MATERIALS

Various nonlinear optical (NLO) techniques have been used to probe intramolecular interactions in solar cell materials. This has included measurements of two-photon absorption with both nanosecond and femtosecond pulses via either nonlinear transmission measurements or through Z-scan measurements.[123–126] In either case, both solution and thin film samples have been investigated.[127,128] It was with these experiments that it was discovered that certain organic structures could demonstrate enhanced intramolecular interactions in comparison to their linear analogs. It was later suggested that various materials with strong two-photon cross sections might have stronger electronic coupling between units of the macromolecule which might lead to longer exciton diffusion lengths. Thus, larger two-photon cross-section could be connected with larger photovoltaic efficiency. The explanation for this enhancement was based on time-resolved measurements with model compounds to probe the relative strengths of interactions. For example, NLO measurements were the critical tool used in discovering that certain multichromophore systems may exhibit very strong optical limiting effects with nanosecond pulses in the visible and possible in the infrared spectral regions.[129] These measurements have also proven useful in characterizing the increase of NLO effects in new multichromophore structures. More detailed NLO measurements have also been carried out to investigate the real and imaginary components of the nonlinear susceptibility by use of degenerated four-wave mixing (DFWM) spectroscopy (Figure 3.14).[130–132] The complete experimental apparatus used for probing the third-order NLO effects in organic PV materials has been discussed extensively in the literature.[133] To obtain two-photon absorption cross sections, one usually follows the two-photon excited fluorescence (TPEF) method.[134] The laser used for these studies is typically 700 to 980 nm fs diode-pumped mode-locked Ti–sapphire laser which can be coupled to an OPO for greater wavelength selection (Figure 3.15).[134]

This light excites the two-photon transition in the molecule and the fluorescence is collected into a monochromator. The output from the monochromator is coupled to a photomultiplier tube. The photons are converted into counts by a photon-counting unit. A logarithmic plot between collected fluorescence photons and input intensity gives a slope of two, ensuring a quadratic dependence. The intercept allows one to calculate the two-photon absorption cross sections at different wavelengths. Measurements of two-photon absorption cross sections in organic solar cell macro-molecules have resulted in new information regarding the specific synthetic design criteria for the fabrication of superior hole and electron transporters for solar device applications. For example, Figure 3.15 shows a system of organic macromolecular materials that can be evaluated by the two-photon fluorescence method in order to gain insight into the degree of electronic coupling as well as the possible energy

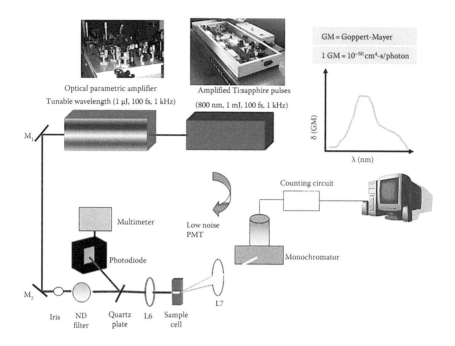

FIGURE 3.14 **(See color insert.)** The two-photon excited fluorescence experimental apparatus.

transfer pathways in a material. A system of thiophene dendrimers was investigated in order to evaluate the delocalization length in these systems.[133] Longer delocalization lengths could have very positive effects on the materials' ability to absorb solar energy and transfer the absorbed energy to different parts of the device such as to the interface.[135,136]

From the important measurements described in this chapter, it is hoped that the reader appreciates the rather extensive detailed knowledge that is required in designing new solar cell materials and device structures. Indeed, the initial discussion of this chapter regarding the measurement of the solar cell device efficiency is an important step in the discovery process. The guidance of experts in this area of standardization of solar cell efficiency measurements has allowed for a fairly straightforward approach to report the detailed parameters of a particular device. The apparatus and materials needed to investigate the efficiency of a device are readily available at this point. Also, there are resources in the United States to have particular devices tested and compared against other standardized devices. It is still hoped that the standardization process will continue to grow in popularity for those working in this field. Equally as important, the measurements of the materials to be used in organic solar cell devices require a multidisciplinary approach of scientists and engineers with diverse interests. The measurement of the surface potential as well as the mobility of electrons is a very detailed and important experiment that can explain the variation of electronic fill factor, V_{oc}, of a material in the device. Measurements such as Kelvin probe spectroscopy have made an impact on the discussion of the importance

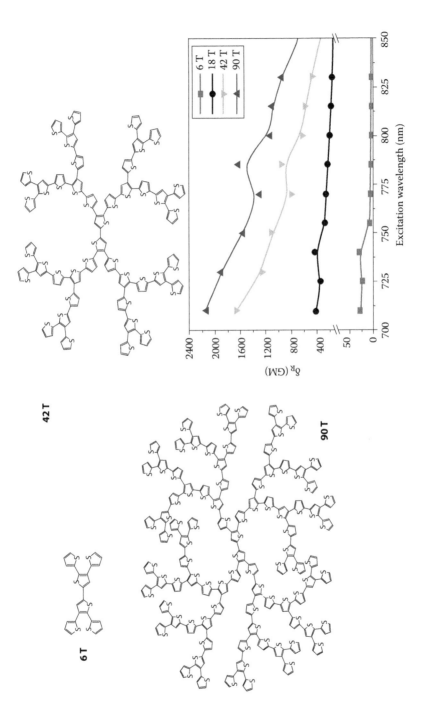

FIGURE 3.15 Two-photon measurements can give insight into intramolecular interactions important to solar cell materials properties. (From Harpham, M.R. et al., *J. Am. Chem. Soc.*, 131, 973, 2009.)

of interfaces. The use of optical spectroscopy such as time-resolved fluorescence and absorption as well as two-photon absorption has been critical to our understanding of the diffusion of excitons in organic materials useful in solar cell devices. New techniques are evolving into the discussion of materials' structure–function relationships as a result of spectroscopic measurements. The key is to find new measurements that can focus on specific parameters important to the operation of an organic solar cell in the vast background of many processes.

REFERENCES

1. Editorial, Solar cell woes, *Nature Photonics*, 8, 665, 2014.
2. Editorial, Bringing solar cell efficiencies into the light, *Nature Nanotechnology*, 9, 657, 2014.
3. K.A. Emery, Measurement and characterization of solar cells and modules, *Handbook of Photovoltaic Science and Engineering*, 2nd edn., Chapter 18, pp. 797–840, A. Luque and S. Hegedus (eds.), John Wiley & Sons, West Sussex, U.K., 2011.
4. NREL, Certificate Number 1239.02 ISO 17025 accredited for photovoltaic secondary cell, secondary module and primary reference cell calibration, American Association for Laboratory Accreditation (A2LA), Golden, CO.
5. (a) K. Emery and H. Field, Artificial enhancements and reductions in the PV efficiency, *Proceedings of 24th IEEE Photovoltaic Specialists Conference*, Waikoloa, HI, p. 1833, 1994; (b) K. Emery, D. Dunlavy, H. Field, and T. Moriarty, Photovoltaic spectral responsivity measurements, *Proceedings of Second World Conference and Exhibition on Photovoltaic Solar Energy Conversion*, Waikoloa, HI, Joint Research Center report EUR 18656, p. 2298, 1998; (c) T. Moriarty and K. Emery, Thermophotovoltaic cell temperature measurement issues, *Proceedings of Fourth NREL TPV Conference*, Waikoloa, HI, AIP Proceedings 460, p. 301, 1998; (d) C. Osterwald and K. Emery, Spectroradiometric sun photometry, *Journal of Atmospheric and Oceanic Technology*, 17, 1171, 2000; (e) K. Emery, D. Myers, and S. Kurtz, What is the appropriate reference spectrum for characterizing concentrator cells?, *Proceedings of 29th IEEE Photovoltaic Specialists Conference*, Waikoloa, HI, pp. 840–843, 2002; (f) K. Emery, Uncertainty analysis of certified photovoltaic measurements at the National Renewable Energy Laboratory, NREL technical report NREL/TP-520-45299, August 2009.
6. V. Shrotriya, G. Li, Y. Yao, T. Moriarty, K. Emery, and Y. Yang, Accurate measurement and characterization of organic solar cells, *Advanced Functional Materials*, 16(15), 2016–2023, 2006.
7. J. Nelson, *The Physics of Solar Cells*, Imperial College Press, London, U.K., 2003.
8. IEC, Photovoltaic devices: Part 3: Measurement principles for terrestrial photovoltaic (PV) solar devices with reference spectral irradiance data, IEC 904-3, International Electro Technical Commission, Geneva, Switzerland, 1989.
9. ASTM, Standard tables for reference solar spectral irradiance at air mass 1.5: Direct normal and hemispherical for a 37° tilted surface, Standard G159-99, American Society for Testing and Materials, West Conshohocken, PA, 1999.
10. ASTM G173-03, 2012, Standard tables for reference solar spectral irradiances: Direct normal and hemispherical on 37° tilted surface, ASTM International, West Conshohocken, PA, 2012. http://www.astm.org/Standards/G173.htm. Accessed February 17, 2017.
11. C.R. Osterwald, M.W. Wanlass, T. Moriarty, M.A. Steiner, and K.A. Emery, Effects of spectral error in efficiency measurements of GaInAs-based concentrator solar cells, NREL Technical report NREL/TP05200-60748, NREL operated by the Alliance for Energy Sustainability, LLC, March 2014.

12. K.A. Emery, C.R. Osterwald, T.W. Cannon, D.R. Myers, J. Burdick, T. Glatfelter, W. Czubatyj, and J. Yang, Methods for measuring solar cell efficiency independent of reference cell or light source, *IEEE PV Specialists Conference Proceedings*, Las Vegas, NV, pp. 623–628, 1985.

13. K. Emery, D. Dunlavy, H. Field, and T. Moriarty, Photovoltaic spectral responsivity measurements, *Second World Conference and Exhibition on Photovoltaic Solar Energy Conversion*, pp. 6–10, Vienna, Austria, July 1998.

14. S. Nann and K. Emery, Spectral effects on PV-device rating, *Solar Energy Materials & Solar Cells*, 27(3), 189, 1992.

15. R.W. Andrews, A. Pollard, and J.M. Pearce, Improved parametric empirical determination of module short circuit current for modelling and optimization of solar photovoltaic systems, *Solar Energy*, 86(9), 2240, 2012.

16. C.R. Osterwald, Translation of device performance measurements to reference conditions, *Solar Cells*, 18(3), 269, 1986.

17. WMO Measurement of radiation, *Guide to Meteorological Instruments and Methods of Observation*, Chapter 7, M. Jarraud (ed.), World Meteorological Organization, Geneva, Switzerland, WMO-No.8, 2008.

18. K. Emery, Solar simulators and I–V measurement methods/device performance, *Solar Cells*, 18(3–4), 251, 1986.

19. J.M. Olson, Simulation of nonuniform irradiance in multijunction IIIV solar cells, *IEEE Photovoltaic Specialists Conference (PVSC)*, Honolulu, HI, p. 201, 2010.

20. Standard specification for direct normal spectrum solar simulation for terrestrial photovoltaic testing, ASTM standard E928, revised 1985, American Society for Testing and Materials, Philadelphia, PA, 1983.

21. NASA, Terrestrial photo voltaic measurement procedures, TM73702, ERDA/NASA/1022-77/16, June 1977.

22. M.K. Chawla, A step by step guide to selecting the 'right' Solar Simulator for your solar cell testing application, Photo Emission Tech application note, 2011, www.photoemission.com/techpapers.

23. A. Luque and S. Hegedus, *Handbook of Photovoltaic Science and Engineering*, John Wiley & Sons, Hoboken, NJ, March 29, 2011.

24. M.D. Redwood, R. Dhillon, R.L. Orozco, X. Zhang, D.J. Binks, M. Dickinson, and L.E. Macaskie, Enhanced photosynthetic output via dichroic beam-sharing, *Biotechnology Letters*, 34(12), 2229–2234, 2012.

25. M. Osborne, Konica Minolta's AK Series photovoltaic reference cells provide consistent cell measurements, PV Tech, London, U.K., 2011, http://www.pv-tech.org/product_reviews/konica_minoltas_ak_series_photovoltaic_reference_cells_provide_consistent_c.

26. L. Dunn and M. Gostein, Light soaking measurements of commercially available CIGS PV modules, *Proceedings of the 38th IEEE Photovoltaic Specialists Conference*, Austin, TX, June 3–8, 2012.

27. M. Gostein and L. Dunn, Light soaking effects on photovoltaic modules: Overview and literature review, *Proceedings of the 37th IEEE Photovoltaics Specialists Conference*, Seattle, WA, June 19–24, 2011.

28. IEC System for Conformity Testing and Certification of Electrical Equipment, Thin-film terrestrial photovoltaic (PV) modules—Design qualification and type approval IEC 61646, 1st edn., UL Group, Northbrook, IL, 1998.

29. L.J. Rozanski, C.T.G. Smith, K.K. Gandhi, M.J. Beliatis, G.D.M.R. Dabera, G. Dabera, K.D.G.I. Jayawardena, A.A.D.T. Adikaari, M.J. Kearney, and S.R.P. Silva, A critical look at organic photovoltaic fabrication methodology: Defining performance enhancement parameters relative to active area, *Solar Energy Materials & Solar Cells*, 130, 513, 2014.

30. K. Nishiokaa, T. Hatayamaa, Y. Uraokaa, T. Fuyukia, R. Hagiharab, and M. Watanabe, Field-test analysis of PV system output characteristics focusing on module temperature, *Solar Energy Materials & Solar Cells*, 75, 665, 2003.

31. J.F. Mendes, P. Horta, M.J. Carvalho, and P. Silva, Solar thermal collectors in polymeric materials: A novel approach towards higher operating temperatures, *Proceedings of ISES World Congress*, 32, 640, 2009.

32. N. Yue, Planar packaging and electrical characterization of high temperature SiC power electronic devices, Master of Science in Materials Science and Engineering thesis submitted to the Faculty of the Virginia Polytechnic Institute and State University, 2008.

33. J. Kettle, R.M. Perks, and P. Dunstan, Localised joule heating in AlGaInP light emitting diodes, *Electronic Letters*, 42(19), 1122, 2006.

34. R. Roesch, K.R. Eberhardt, S. Engmann, G. Gobsch, and H. Hoppe, Polymer solar cells with enhanced lifetime by improved electrode stability and sealing, *Solar Energy Materials & Solar Cells*, 117, 59, 2013.

35. H. Hoppe, T. Glatzel, M. Niggemann, A. Hinsch, M.C. Lux-Steiner, and N.S. Sariciftci, Kelvin probe force microscopy study on conjugated polymer/fullerene bulk heterojunction organic solar cells, *Nano Letters*, 5(2), 269, 2005.

36. N. Hayashi, H. Ishii, Y. Ouchi, and K. Seki, Examination of band bending at buckminsterfullerene (C-60)/metal interfaces by the Kelvin probe method, *Journal of Applied Physics*, 92(7), 3784, 2002.

37. M. Nonnenmacher, M.P. Oboyle, and H.K. Wickramasinghe, Kelvin probe force microscopy, *Applied Physics Letters*, 58(25), 2921, 1991.

38. L. Kronik and Y. Shapira, Surface photovoltage phenomena: Theory, experiment, and applications, *Surface Science Reports*, 37(1), 206, 1999.

39. a) H.O. Jacobs, P. Leuchtmann, O.J. Homan, and A. Stemmer, Resolution and contrast in Kelvin probe force microscopy, *Journal of Applied Physics*, 84(3), 1168, 1998. b) W. Melitz, J. Shen, A. Kummel, S. Lee, Kelvin probe force microscopy and its application, *Surface Science Reports*, 66, 1–27, 2011.

40. G.Y. Liu, S. Xu, and Y.L. Qian, Nanofabrication of self-assembled monolayers using scanning probe lithography, *Accounts of Chemical Research*, 33(7), 457, 2000.

41. V. Palermo, M. Palma, and P. Samori, Electronic characterization of organic thin films by Kelvin probe force microscopy, *Advanced Materials*, 18(2), 145, 2006.

42. M.-C. Wu, H.-C. Liao, Y.-C. Cho, C.-P. Hsu, T.-H. Lin, W.-F. Su, A. Sápi et al., Photocatalytic activity of nitrogen-doped TiO_2-based nanowires: A photo-assisted Kelvin probe force microscopy study, *Journal of Nanoparticle Research*, 16, 2143, 2014.

43. Y.Y. Lin, T.H. Chu, S.S. Li, C.H. Chuang, C.H. Chang, W.F. Su, C.P. Chang, M.W. Chu, and C.W. Chen, Interfacial nanostructuring on the performance of polymer/TiO_2 nanorod bulk heterojunction solar cells, *Journal of American Chemical Society*, 131(10), 3644, 2009.

44. T. Glatzel, H. Hoppe, N.S. Sariciftci, M.C. Lux-Steiner, and M. Komiyama, Kelvin probe force microscopy study of conjugated polymer/fullerene organic solar cells, *Japanese Journal of Applied Physics Part 1*, 44(7B), 5370, 2005.

45. J.C. Scott, Metal–organic interface and charge injection in organic electronic devices, *Journal of Vacuum Science and Technology*, 21(3), 521, 2003.

46. A. Liscio, G. De Luca, F. Nolde, V. Palermo, K. Muellen, and P. Samori, Photovoltaic charge generation visualized at the nanoscale: A proof of principle, *Journal of American Chemical Society*, 130(3), 780, 2008.

47. R. Asahi, T. Morikawa, T. Ohwaki, K. Aoki, and Y. Taga, Visible-light photocatalysis in nitrogen-doped titanium oxides, *Science*, 293(5528), 269, 2001.

48. M. Beu, K. Klinkmuller, and D. Schlettwein, Use of Kelvin probe force microscopy to achieve a locally and time-resolved analysis of the photovoltage generated in dye-sensitized ZnO electrodes, *Physica Status Solidi, A*, 211(9), 1960, 2014.

49. F. Fabregat-Santiago, J. Bisquert, G. Garcia-Belmonte, G. Boschloo, and A. Hagfeldt, Influence of electrolyte in transport and recombination in dye-sensitized solar cells studied by impedance spectroscopy, *Solar Energy Materials & Solar Cells*, 87(1), 117, 2005.

50. F. Fabregat-Santiago, G. Garcia-Belmonte, I. Mora-Sero, and J. Bisquert, Characterization of nanostructured hybrid and organic solar cells by impedance spectroscopy, *Physical Chemistry Chemical Physics*, 13(20), 9083, 2011.

51. G. Garcia-Belmonte, A. Munar, E.M. Barea, J. Bisquert, I. Ugarte, and R. Pacios, Charge carrier mobility and lifetime of organic bulk heterojunctions analyzed by impedance spectroscopy, *Organic Electronics*, 9(5), 847, 2008.

52. B.Y. Chang and S.M. Park, Electrochemical impedance spectroscopy, *Annual Reviews of Analytical Chemistry*, 3, 207, 2010.

53. M. Adachi, M. Sakamoto, J.T. Jiu, Y. Ogata, and S. Isoda, Determination of parameters of electron transport in dye-sensitized solar cells using electrochemical impedance spectroscopy, *Journal of Physical Chemistry B*, 110(28), 13872, 2006.

54. W.H. Leng, Z. Zhang, J.Q. Zhang, and C.N. Cao, Investigation of the kinetics of a TiO_2 photoelectrocatalytic reaction involving charge transfer and recombination through surface states by electrochemical impedance spectroscopy, *Journal of Physical Chemistry B*, 109(31), 150008, 2005.

55. M.M. Musiani, Characterization of electroactive polymer layers by electrochemical impedance spectroscopy, *Electrochimica Acta*, 35(10), 1665, 1999.

56. M.A. Vorotyntsev, J.P. Badiali, and G. Inzelt, Electrochemical impedance spectroscopy of thin films with two mobile charge carriers: Effects of the interfacial charging, *Journal of Electroanalytical Chemistry*, 4721, 7, 1999.

57. G. Garcia-Belmonte, P.P. Boix, J. Bisquert, M. Sessolo, and H.J. Bolink, Simultaneous determination of carrier lifetime and electron density-of-states in P3HT:PCBM organic solar cells under illumination by impedance spectroscopy, *Solar Energy Materials & Solar Cells*, 94(2), 366, 2010.

58. A. Guerrero, B. Dorling, T. Ripolles-Sanchis, M. Aghamohammadi, E. Barrena, M. Campoy-Quiles, and G. Garcia-Belmonte, Interplay between fullerene surface coverage and contact selectivity of cathode interfaces in organic solar cells, *ACS Nano*, 7(5), 4637, 2013.

59. A. Guerrero, S. Loser, G. Garcia-Belmonte, C.J. Bruns, J. Smith, H. Miyauchi, S.I. Stupp, J. Bisquert, and T.J. Marks, Solution-processed small molecule:fullerene bulk-heterojunction solar cells: Impedance spectroscopy deduced bulk and interfacial limits to fill-factors, *Physical Chemistry Chemical Physics*, 15(39), 16456, 2013.

60. C.M. Proctor, J.A. Love, and T.Q. Nguyen, Mobility guidelines for high fill factor solution-processed small molecule solar cells, *Advanced Materials*, 26(34), 5957, 2014.

61. K.M. Coakley and M.D. McGehee, Conjugated polymer photovoltaic cells, *Chemistry of Materials*, 16(23), 4533, 2004.

62. S.R. Scully and M.D. McGehee, Effects of optical interference and energy transfer on exciton diffusion length measurements in organic semiconductors, *Journal of Applied Physics*, 100(3), 034907, 2006.

63. A. Haugeneder, M. Neges, C. Kallinger, W. Spirkl, U. Lemmer, J. Feldmann, U. Scherf, E. Harth, A. Gugel, and K. Mullen, Exciton diffusion and dissociation in conjugated polymer fullerene blends and heterostructures, *Physical Review B*, 59(23), 15346, 1999.

64. J.J.M. Halls, K. Pichler, R.H. Friend, S.C. Moratti, and A.B. Holmes, Exciton diffusion and dissociation in a poly(p-phenylenevinylene)/C-60 heterojunction photovoltaic cell, *Applied Physics Letters*, 68(22), 3120, 1996.

65. M. Yan, L.J. Rothberg, F. Papadimitrakopoulos, M.E. Galvin, and T.M. Miller, Spatially indirect excitons as primary photoexcitations in conjugated polymers, *Physical Review Letters*, 72(7), 1104, 1994.

66. B.A. Gregg, J. Sprague, and M.W. Peterson, Long-range singlet energy transfer in per-ylene bis(phenethylimide) films, *Journal of Physical Chemistry B*, 101(27), 5362, 1997.

67. D.R. Haynes, A. Tokmakoff, and S.M. George, Distance dependence of electronic energy transfer between donor and acceptor layers-p-terphenyl and 9,10 diphenylan-tathracene, *Journal of Chemical Physics*, 100(3), 1968, 1994.

68. H. Kuhn, Classical aspects of energy transfer in molecular systems, *Journal of Chemical Physics*, 53(1), 101, 1970.

69. Y.X. Liu, M.A. Summers, C. Edder, J.M.J. Frechet, and M.D. McGehee, Using reso-nance energy transfer to improve exciton harvesting in organic-inorganic hybrid photo-voltaic cells, *Advanced Materials*, 17(24), 2960, 2006.

70. W.A. Luhman and R.J. Holmes, Investigation of energy transfer in organic photovol-taic cells and impact on exciton diffusion length measurements, *Advanced Functional Materials*, 21, 764, 2011.

71. D.E. VandenBout, W.T. Yip, D.H. Hu, D.K. Fu, T.M. Swager, and P.F. Barbara, Discrete intensity jumps and intramolecular electronic energy transfer in the spectroscopy of single conjugated polymer molecules, *Science*, 277, 5329, 1997.

72. J. Yu, D.H. Hu, and P.F. Barbara, Unmasking electronic energy transfer of conjugated polymers by suppression of O(2) quenching, *Science*, 289, 5483, 2000.

73. H. Fidder, J. Knoester, and D.A. Wiersma, Observation of one exciton to two exciton transition in a J aggregate, *Journal of Chemical Physics*, 98(8), 6564, 1993.

74. S. Tretiak, V. Chernyak, and S. Mukamel, Localized electronic excitations in phenyl-acetylene dendrimers, *Journal of Physical Chemistry B*, 102(18), 3310, 1998.

75. J. Clark, C. Silva, R.H. Friend, and F.C. Spano, Role of intermolecular coupling in the photophysics of disordered organic semiconductors: Aggregate emission in regioregular polythiophene, *Physical Review Letters*, 98(20), 206406, 2007.

76. V. Coropceanu, J. Cornil, D.A. da Silva, Y. Olivier, R. Silbey, and J.L. Bredas, Charge transport in organic semiconductors, *Chemical Reviews*, 107(4), 926, 2007.

77. P. Peumans, A. Yakimov, and S.R. Forrest, Small molecular weight organic thin-film photodetectors and solar cells, *Journal of Applied Physics*, 93(7), 3693, 2003.

78. D.F. O'Brien, M.A. Baldo, M.E. Thompson, and S.R. Forrest, Improved energy transfer in electrophosphorescent devices, *Applied Physics Letters*, 74(3), 442, 1999.

79. K. Celebi, T.D. Heidel, and M.A. Baldo, Simplified calculation of dipole energy transport in a multilayer stack using dyadic Green's functions, *Optics Express*, 15(4), 1762, 2007.

80. P.V. Kamat, Meeting the clean energy demand: Nanostructure architectures for solar energy conversion, *Journal of Physical Chemistry C*, 111(7), 2834, 2007.

81. D.R. Kozub, K. Vakhshouri, S.V. Kesava, C. Wang, A. Hexemer, and E.D. Gomez, Direct measurements of exciton diffusion length limitations on organic solar cell perfor-mance, *Chemical Communications*, 48(47), 5859, 2012.

82. L.H. Jimison, M.F. Toney, I. McCulloch, M. Heeney, and A. Salleo, Charge-transport anisotropy due to grain boundaries in directionally crystallized thin films of regioregular poly(3-hexylthiophene), *Advanced Materials*, 21(16), 1568, 2009.

83. K. Vakhshouri, S.V. Kesava, D.R. Kozub, and E.D. Gomez, Characterization of the mesoscopic structure in the photoactive layer of organic solar cells: A focused review, *Materials Letters*, 90, 97, 2013.

84. G.D. Sharma, G.E. Zervaki, P.A. Angaridis, T.N. Kitsopoulos, and A.G. Coutsolelos, Triazine-bridged porphyrin triad as electron donor for solution-processed bulk hetero-junction organic solar cells, *Journal of Physical Chemistry C*, 118(11), 5968, 2014.

85. D.S. Tyson and F.N. Castellano, Intramolecular singlet and triplet energy transfer in a ruthenium(II) diimine complex containing multiple pyrenyl chromophores, *Journal of Physical Chemistry A*, 03(50), 10955, 1999.

86. C.C. Jumper, J.M. Anna, A. Stradomska, J. Schins, M. Myahkostupov, V. Prusakova, D.G. Oblinsky, F.N. Castellano, J. Knoester, and G.D. Scholes, Intramolecular radiationless transitions dominate exciton relaxation dynamics, *Chemical Physics Letters*, 599, 23, 2014.

87. J.W. Cho, H. Yoo, J.E. Lee, Q.F. Yan, D.H. Zhao, and D. Kim, Intramolecular interactions of highly π-conjugated perylenediimide oligomers probed by single-molecule spectroscopy, *Journal of Physical Chemistry Letters*, 5(21), 3895, 2014.

88. J. Klafter and R. Silbey, Derivation of the continuous time random walk equation, *Physical Review Letters*, 44(2), 55, 1980.

89. A. BarHaim, J. Klafter, and R. Kopelman, Dendrimers as controlled artificial energy antennae, *Journal of American Chemical Society*, 19(26), 6197, 1997.

90. S. Tretiak and S. Mukamel, Density matrix analysis and simulation of electronic excitations in conjugated and aggregated molecules, *Chemical Reviews*, 102(9), 3172, 2002.

91. P.V. Kamat, Photochemistry on nonreactive and reactive semiconductor surfaces, *Chemical Reviews*, 93(1), 267, 1993.

92. W.D. Zeng, Y.M. Cao, Y. Bai, Y.H. Wang, Y.S. Shi, M. Zhang, F.F. Wang, C.Y. Pan, and P. Wang, Efficient dye-sensitized solar cells with an organic photosensitizer featuring orderly conjugated ethylenedioxythiophene and dithienosilole blocks, *Chemistry of Materials*, 22(5), 1915, 2010.

93. H. Ohkita, S. Cook, Y. Astuti, W. Duffy, S. Tierney, W. Zhang, M. Heeney et al., Charge carrier formation in polythiophene/fullerene blend films studied by transient absorption spectroscopy, *Journal of American Chemical Society*, 130(10), 3030, 2008.

94. J.J. He, G. Benko, F. Korodi, T. Polivka, R. Lomoth, B. Akermark, L.C. Sun, A. Hagfeldt, and V. Sundstrom, Modified phthalocyanines for efficient near-IR sensitization of nanostructured TiO_2 electrode, *Journal of American Chemical Society*, 124(17), 4922, 2002.

95. N.A. Anderson and T.Q. Lian, Ultrafast electron transfer at the molecule-semiconductor nanoparticle interface, *Annual Review of Physical Chemistry*, 56, 491, 2005.

96. R.D. Pensack and J.B. Asbury, Beyond the adiabatic limit: Charge photogeneration in organic photovoltaic materials, *Journal of Physical Chemistry Letters*, 1(15), 2255, 2010.

97. J.L. Bredas, J.E. Norton, J. Cornil, and V. Coropceanu, Molecular understanding of organic solar cells: The challenges, *Accounts of Chemical Research*, 42(11), 1691, 2009.

98. J. Zhang, M.K.R. Fischer, P. Bauerle, and T. Goodson, Energy migration in dendritic oligothiophene-perylene bisimides, *Journal of Physical Chemistry B*, 117(16), 4204, 2013.

99. G. Ramakrishna, A. Bhaskar, and T. Goodson, Ultrafast excited state relaxation dynamics of branched donor-pi-acceptor chromophore: Evidence of a charge-delocalized state, *Journal of Physical Chemistry B*, 110(42), 20872, 2006.

100. G. Ramakrishna, T. Goodson, J.E. Rogers-Haley, T.M. Cooper, D.G. McLean, and A. Urbas, Ultrafast intersystem crossing: Excited state dynamics of platinum acetylide complexes, *Journal of Physical Chemistry C*, 113(3), 1060, 2009.

101. O. Varnavski, G. Menkir, T. Goodson III, and P.L. Burn, Ultrafast polarized fluorescence dynamics in an organic dendrimer, *Applied Physics Letters*, 77, 1120–1122, 2000.

102. O.P. Varnavski, L. Sukhomlinova, R. Tweig, G.C. Bazan, and T. Goodson III, Coherent effects in energy transport in model dendritic structures investigated by ultra-fast fluorescence anisotropy spectroscopy, *Journal of American Chemical Society*, 124, 1736–1743, 2002.

103. O. Varnavski, I.D.W. Samuel, L.O. Palsson, R. Beavington, P.L. Burn, and T. Goodson III, Investigations of excitation energy transfer and intramolecular interactions in a nitrogen corded distrylbenzene dendrimer system, *Journal of Chemical Physics*, 116, 8893–8903, 2002.

104. M.I. Ranasinghe, O.P. Varnavski, J. Pawlas, S.I. Hauck, J. Louie, J.F. Hartwig, and T. Goodson III, Femtosecond energy transport in triarylamine dendrimers, *Journal of American Chemical Society*, 124, 6520–6521, 2002.

105. M. Ranasinghe, Y. Wang, and T. Goodson III, Dynamics of energy transport in organic dendrimers at low (4K) temperature, *Journal of American Chemical Society*, 125(18), 5258–5259, 2003.

106. Y. Wang, M. Ranasinghe, and T. Goodson III, Mechanistic studies of energy transport in a phosphorous cored branching structure, *Journal of American Chemical Society*, 125, 9562–9563, 2003.

107. R. West, S. Lahankhar, H.B. Xie, O. Varnavski, M. Ranasinghe, and T. Goodson III, Electronic coupling in organic branched macromolecules investigated by nonlinear optical and fluorescence upconversion spectroscopy, *Journal of Chemical Physical*, 120(1), 337–344, 2004.

108. O. Varnavski and T. Goodson III, Exciton dynamics in a branched molecule probed with three pulse photon echo peak shift and transient grating spectroscopy, *Journal of Physical Chemistry*, 108(29), 10484–10492, 2004.

109. M. Ransinghe, M.W. Hager, C.B. Gorman, and T. Goodson III, Energy transfer in phenylacetylene dendrimers revisited, *Journal of Physical Chemistry*, 108, 8543–8549, 2004.

110. T. Goodson III, Optical excitations in novel organic branched structures investigated by time-resolved and nonlinear optical spectroscopy, *Accounts of Chemical Research*, 38, 99–107, 2005.

111. T. Goodson III, Photochemistry of dendrimers, *Annual Review of Physical Chemistry*, 56, 581–603, 2005.

112. X. Yan, J. Pawlas, J. Hartwig, and T. Goodson III, Polaron delocalization in novel organic ladder macromolecules, *Journal of American Chemical Society*, 127(25), 9105–9116, 2005.

113. A. Bhaskar, R. Guda, M.M. Haley, and T.G. Goodson III, Building symmetric two-dimensional two-photon materials, *Journal of American Chemical Society*, 128(43), 13972–13973, 2006.

114. M. Guo, O. Varnavski, A. Narayanan, O. Mongin, J.P. Majoral, M. Blanchard-Desce, and T. Goodson III, Investigations of energy migration in an organic dendrimer macromolecule for sensory signal amplification, *Journal of Physical Chemistry A*, 113(16), 4763–4771, 2009.

115. D. Flynn, G. Ramakrishna, H.B. Yang, B. Northrop, P. Stang, and T. Goodson III, Ultrafast optical excitations in supramoleculars metallacycles with charge transfer properties, *Journal of American Chemical Society*, 132, 1348–1358, 2010.

116. J.E. Donehue, O.P. Varnavski, R. Chemborski, M. Lyoda, and T. Goodson III, Probing coherence in synthetic cyclic light-harvesting pigments, *Journal of American Chemical Society*, 133, 4819–4828, 2011.

117. T. Goodson III, Self-assembled organic aggregates, *Journal of Physical Chemistry Letters*, 2(8), 932–933, 2011.

118. T. Goodson III, Understanding the intricacies of organic photovoltaics, *Journal of Physical Chemistry Letters*, 24, 3146–3146, 2011.

119. J. Furgal, J. Jung, T. Goodson III, and R. Laine, Analyzing structure-photophysical property relationships for isolated T-8, T-10, and T-12 stilbenevinylsilesquioxanes, *Journal of American Chemical Society*, 135, 12259–12269, 2013.

120. T. Goodson III, Recent advances in designing molecular assemblies, *Journal of Physical Chemistry Letters*, 4, 2705–2706, 2013.

121. O. Varnavski, A. Leanov, L. Liu, J. Takacs, and T. Goodson III, Large nonlinear refraction and higher order nonlinear optical effects in a novel organic dendrimer, *Journal of Physical Chemistry*, 104, 179–188, 2000.

122. R.G. Ispasoiu, L. Balogh, O.P. Varnavski, D.A. Tomalia, and T. Goodson III, Large opti-cal limiting and ultrafast luminescence dynamics from novel metal-dendrimer nano-composite materials, *Journal of American Chemical Society*, 122, 11005–11006, 2000.

123. T.E.O. Screen, K.B. Lawton, G.S. Wilson, N. Dolney, R. Ispasoiu, T. Goodson III, S.J. Martin, D.D.C. Bradley, and H.L. Anderson, Synthesis and third order nonlinear optics of a new soluble conjugated porphyrin polymer, *Journal of Material Chemistry*, 11, 312–320, 2001.

124. R. West, Y. Wang, and T. Goodson III, Nonlinear transmission investigations in gold nanostructured materials, *Journal of Physical Chemistry*, 107(15), 3419–3426, 2003.

125. J.E. Raymond, A. Bhaskar, T. Goodson III, N. Makiuchi, K. Ogawa, and Y. Kobuke, Synthesis and two-photon absorption enhancement of porphyrin macrocycles, *Journal of American Chemical Society*, 130(51), 17212–17213, 2008.

126. M. Guo, O. Varnavski, A. Narayan, O. Mongin, J.P. Majoral, M. Blanchard-Desce, and T. Goodson, Investigations of energy migration in an organic dendrimer macromolecule for sensory signal amplification, *Journal of Physical Chemistry A*, 113(16), 4763–4771, 2009.

127. C. Wang, T. Ren, and T. Goodson III, Nonlinear optical properties of Ru-complexes, *Organometallics*, 24(13), 3247–3254, 2005.

128. Y. Wang, X.B. Xie, and T. Goodson III, Enhanced third order NLO properties in den-drimer metal nanocomposites, *Nano Letters*, 5, 2379–2384, 2005.

129. T. Zheng, Z. Cai, R. Ho-Wu, S.H. Yau, V. Shaparov, T. Goodson, and L. Yu, Synthesis of ladder-type thienoacenes and their electronic and optical properties, *Journal of the American Chemical Society*, 138(3), 868–875, 2016.

130. A. Bhaskar, G. Ramakrishna, R. Twieg, and T. Goodson III, Investigations of two-photon absorption properties in branched alkene and alkyne chromophores, *Journal of American Chemical Society*, 128(36), 11840–11849, 2006.

131. O. Varnavski, X. Yan, O. Mongin, M. Blanchard-Desce, and T. Goodson III, Strongly interacting organic conjugated dendrimers with enhanced two-photon absorption, *Journal of Physical Chemistry C*, 111(1), 149–162, 2007.

132. A. Narayan, O. Varnavski, T.M. Swager, and T. Goodson III, Multiphoton fluorescence quenching of conjugated polymers for TNT detection, *Journal of Physical Chemistry C (Letter)*, 112, 881–884, 2007.

133. M.R. Harpham, O. Suzer, Ch.Q. Ma, P. Bauerle, and T. Goodson III, Thiophene den-drimers as entangled photo sensor materials, *Journal of American Chemical Society*, 131, 973–979, 2009.

134. A. Guzman, M.R. Harpham, I. Suzer, M.M. Haley, and T. Goodson III, Spatial control of entangled two-photon absorption with organic chromophores, *Journal of American Chemical Society*, 132(23), 7840–7841, 2010.

135. T.B. Clark, Y. Wang, O. Varnavski, and T. Goodson III, Two-photon and time-resolved fluorescence conformational studies of aggregation in amyloid peptides, *Journal of Physical Chemistry C*, 114(20), 7112–7120, 2010.

136. O.O. Adegoke, I.H. Jung, M. Orr, L.P. Yu, and T. Goodson III, Effect of acceptor strength on optical and electronic properties in conjugated polymers for solar applica-tions, *Journal of the American Chemical Society*, 137(17), 5759, 2015.

4 Organic Solar Cell Architectures Modernized

BASIC CONSTRUCTIONS REVISITED

By now, the reader should realize that there has been a great deal of progress in the science and technology of organic solar cells (Figure 4.1). In particular, the measurements, standardization of methods for testing efficiencies, materials, structure–function relationships, as well as the details of the photo-physics and long-term stability have thoroughly been investigated in a number of promising organic solar materials. Indeed, this is still a critical problem in the sustainability of our world's energy needs and the development of new ideas for the future. While there has been a great deal of talk regarding the future, it is worthy to note that there has also been new and far reaching ideas about organic solar cells, which are not necessarily the norm of what has been discussed in previous chapters. While the commercialization of current organic solar cell technology seems to be finding ground in some markets around the world, there are some that believe that for the case of organic solar there might be a better approach of seriously digging into the depth of newer technology first, a new technology that might offer a world- and game-changing breakthrough in both efficiency and cost.[1] Newer ideas regarding the construction of devices as well as the particular use of materials have been discussed under this context, and this is the focus of this chapter.

Perhaps it should not be surprising that in regard to the development of new ideas and new approaches, one usually looks back to the beginning of this story and first evaluates the initial ideas and suggestions as to why it was originally thought that certain organic systems might really work as solar cell materials. This mainly involves looking at the components of the device and discussing the dynamics of charge carriers in the device. If one thinks about the initial experiments of a photovoltaic effect in an organic crystal, one may suggest that looking at detailed studies of single crystals might be a good starting point to consider. In these experiments, the single crystals are placed between two electrodes and the light is illuminated on one of the sides.[2,3] These experiments led to the conceptual idea that there are more than one exciton dissociation mechanisms, which are important in the charge carrier dynamics. It was then suggested that electron injection into the bulk and hole transport by the organic crystal away from the interface was the main mechanism of the photovoltaic effect in organic crystals.[3] These studies demonstrated the differences between conventional (inorganic) solar cells, which are usually based on silicon or other inorganic semiconductors, and organic solar cells.[4–6] The organic

FIGURE 4.1 (**See color insert.**) Modern residential construction is now seriously considering solar energy in their design criteria. (Credit: www.shutterstock.com, 128612171.)

devices did not appear to operate in the same manner their inorganic counterparts were known to behave. For example, the charge carriers in organic semiconductors are tightly bound to each other in the form of excitons. The excitons only dissociate at interfaces.[6] Reports also demonstrated that in organic solar cells, holes exist primarily in one phase, electrons exist primarily in the other phase, and their movements result directly in current flow.[7] Previous investigations further detailed the importance of the structure of the device (bulk, bilayer, heterojunction) and the device performance.

As mentioned in previous chapters, there have been a great number of lessons learned in the formation of organic electronic devices from the studies of light emitting diodes. As was the case for organic light emitting diodes (OLEDs), for novel organic solar architectures, there was great consideration given to the difficulty in designing new architectures for organic solar cells in terms of the electrodes. While most might conceive that this should be a particularly straightforward issue in devices considering the great deal of previous work on similar devices for OLEDs, the choice and work function of the particular electrodes used can have a tremendous effect on the overall PV efficiency of a device.[8,9] In fact, many believe that certain organic devices may already have the necessary efficiencies to compete with silicon if only they didn't suffer from such large losses associated with the processes and stability of the electrodes.[10] Much of these loss factors relate to the geminate pair and bimolecular recombination losses.[10] In the newer approaches, novel materials for new electrode systems in organic solar cell devices are considered. These electrode systems offer advantages to the commonly used indium tin oxide (ITO) electrodes, which have limitations as found from experience in the field.[11–14] Issues related to reflectivity, reproducibility of the surfaces, and flexibility have caused issues in the past with ITO.[12] Another consideration for new solar cell architectures involves overcoming the Schottky barrier with engineering and materials design. In particular, it is now believed that with particular design and engineering of a particular geometry of a device, one may systematically increase the efficiency of Schottky solar cells above the theoretical limits which were once calculated. The ultimate conversion efficiencies for the new cells are the same as for $p–n$ junction devices. With present technology, improvements of over 50% above the old limits are possible. Indeed, this is a very big direction, as the pay-off could be equally as large. It is believed that because of a Schottky diode's low forward voltage drop, less energy is wasted as heat making them the most efficient choice for applications sensitive to efficiency. For instance, they are used in stand-alone photovoltaic systems to prevent batteries from discharging through the solar panels at night and in grid-connected systems with multiple strings connected in parallel to prevent reverse current flowing from adjacent strings through shaded strings if the bypass diodes have failed. Indeed, if the present technology will expand and improve on the use of Schottky solar devices, it might revolutionize our way of thinking about the limits of solar photon capture and how we might fully utilize this approach for many solar applications. Also, the method in which the devices are constructed and then conditioned may also have a profound effect on the overall efficiency. In the modernization of the organic solar cells, new methods and approaches have been presented to purify and to anneal the material into a form for maximum performance. New reports have started to receive great attention for the increase in performance as a result of these new methodologies. In this chapter we discuss these new approaches and revisit the basic concepts and issues surrounding organic solar cell fabrication.

ESSENTIAL DIAGRAMS AND IDEAS

In considering the basic solar cell device structures, we can see that the earliest organic solar cells were based on single thermally evaporated molecular organic layers sandwiched between two metal electrodes of different work functions as described in previous chapters. The dynamics of the charge carriers can be explained by the formation of a Schottky barrier depending on the insulating behavior of the material between the metal with the lower work function and the p-type organic layer.[15] In many cases a Schottky junction is found to be operative when aluminum or calcium electrodes are used. As mentioned in previous chapters, a major goal in obtaining high efficiencies is to obtain large fill factors (FF). Shown in Figure 4.2 is a circuit diagram for a solar cell. As it can be seen from the figure, one may obtain a larger FF if the shunt resistance is very large to prevent leakage currents and if the series resistance is very low to get a sharp rise in the forward current.[16] There is a summation of the different series resistance contributions in the device.[17] The situation for a Schottky junction is shown in Figure 4.3, for the case of metal contacts. One observes the typical band bending in the depletion region from the Schottky contact. From initial ideas and experiments, it was suggested that this corresponds to an electric field in which excitons can be dissociated.[18]

The theoretical limits of a solar cell were first studied in depth in the 1960s, and these limits are known today as the Detailed Balance Limit or simply the Shockley–Queisser limit.[19–22] The limit describes several loss mechanisms that are inherent

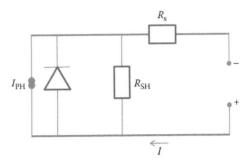

FIGURE 4.2 Circuit diagram of a solar cell.

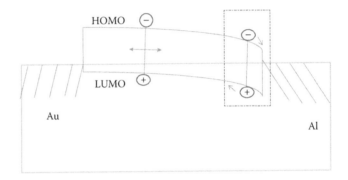

FIGURE 4.3 Schottky contact single-layer device.

to any solar cell design.[19] The first are the losses due to blackbody radiation, a loss mechanism that affects any material object above absolute zero. In the case of solar cells at standard temperature and pressure, this loss accounts for about 7% of the power. The second is an effect known as "recombination," where the electrons created by the photoelectric effect meet the electron holes left behind by previous excitations. In silicon, this accounts for another 10% of the power. However, the dominant loss mechanism is the inability for a solar cell to extract all of the power in the photon, and the associated problem that it cannot extract any power at all from certain photons. This is due to the fact that the electrons have to acquire enough energy to overcome the bandgap of the material and that energy is removed from the energy originally in the photon.[20-22] If the photon has less energy than the bandgap, it is not collected at all. This is a major consideration for conventional solar cells, which are not sensitive to most of the infrared spectrum, although that represents almost half of the power coming from the sun.[23] The opposite is also true in the case of high energy photons, where these photons initially eject an electron to a state high above the bandgap, but this extra energy is lost through collisional relaxation processes. This lost energy turns into heat in the cell, which has the side effect of further increasing blackbody losses. Considering the spectrum losses alone, a solar cell has a peak theoretical efficiency of 48%. Thus, the spectrum losses represent the vast majority of lost power. Including the effects of blackbody radiation and recombination, the efficiency is described by[19]

$$\eta = \frac{qV\phi_s - \phi_r}{\sigma T_{sun}^4} \tag{4.1}$$

where
- q is the electric charge
- V is voltage across the device
- ϕ_s is the incident photon flux entering the device
- ϕ_r is the radiative photon flux leaving the device
- σ is the Stefan–Boltzmann constant
- T_{sun} is the temperature of the sun

A single-junction cell has a theoretical peak performance of about 33.7%. Indeed, this theoretical efficiency presents a major problem to those interested in enhancing the overall efficiency of solar devices. Silicon seems to be tolerant of these losses due to its initially high efficiency. However, this is not the case for many organic devices.

New ideas have considered the Shockley–Queisser limit further and have suggested new avenues toward approaching its illusive milestone. As described earlier, in general, the current–voltage characteristics of organic heterojunctions are often modeled using the generalized Shockley equation derived primarily for inorganic diodes. There is no simple description in the model for organic semiconductor donor–acceptor heterojunction cells. Recently, due to the limited information for organic donor–acceptor devices, there have been extensions of the Shockley–Queisser limit.[24] In this

$$J = J_{sD}\left[\exp\left(\frac{qV_a}{n_D k_b T}\right) - \frac{k_{PPd}}{k_{PPd,eq}}\right] + J_{sA}\exp\left(\frac{qV_a}{n_A k_b T}\right) - \frac{k_{PPd}}{k_{PPd,eq}} - q\eta_{PPd}J_X \tag{4.2}$$

new approach (see Equation 4.2), the current density-voltage characteristic is derived specifically for donor–acceptor heterojunction solar cells, and this model predicts the general dependence of dark current, open-circuit voltage (V_{oc}), and short-circuit current (J_{sc}) on temperature and light intensity as well as the maximum V_{oc} for a given donor–acceptor material pair.[24] The "trap limited ideal diode equation" is similar in its form to the Shockley equation but different in its interpretation based on the difference in excited state description between organic and inorganic materials.[24] It was suggested that in this case, the trap-limited recombination due to disorder at the donor–acceptor interface leads to the introduction of two temperature-dependent ideality factors. Examples were used to demonstrate how this new model may offer a better description were the Shockley equation breaks down. This investigation (and others) concluded that the polaron pair recombination rate is a key factor that determines the J–V characteristics in the dark and under illumination.[24–26] A polaron is a quasi particle often used in device physics discussions to describe the interaction

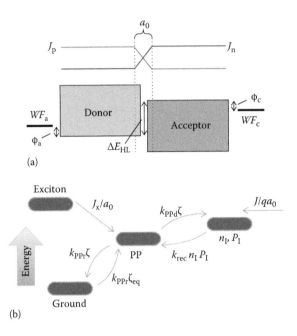

(a)

(b)

FIGURE 4.4 **(See color insert.)** (a) Energy-level diagram showing the anode and cathode work functions. (b) Processes occurring within the HJ region. Excitons diffuse, with current density, J_X, to the HJ and undergo charge transfer to form polaron pairs. These may recombine at rate k_{PPr} or dissociate with rate k_{PPd}. (From Giebink, N.C. et al., *Phys. Rev. B*, 82, 15, 2010.)

between electrons and atoms in a solid material. These quasi particles are excited and their relaxation or recombination rates provide a key factor in the overall efficiency of organic donor–acceptor type solar cells (Figure 4.4).[24–27]

Thus, there appears to be hope for new approaches to explain the rates and efficiencies in organic solar cells and that we may not have to rely on the inorganic descriptions. Even with the new descriptions in hand, however, one must still realize that the exciton diffusion length for most organic solar cell materials is still below ~20 nm.[28–30] Indeed, only those excitons generated in a small region within 20 nm from the contacts contribute to the photocurrent.[29] Due to the high series resistances, these materials show a low FF and a field-dependent charge carrier collection. These thin film devices can work well as photodetectors, as under a high reverse bias, the electric field drives the created charges to the electrodes. The illumination intensity dependence of the short-circuit photocurrent is sub-linear, as expected for bimolecular recombination, as the probability for recombination is proportional to both electron and hole concentrations. As discussed previously, the short exciton diffusion length in many organic systems will affect the overall mobility and efficiency of the device as well as the action of the Schottky barrier. In the modernization of these devices, it has become clear that the use of more layers and more sophistication in connecting the layers is important in improving many aspects of organic solar cell devices. Multijunction solar cells or tandem cells are solar cells containing

FIGURE 1.1 The gods of ancient Egypt—Aten and Ra. Ra in the solar bark. (Credit: www. shutterstock.com, 122013538.)

FIGURE 1.2 Power plant using renewable solar energy from the sun. (Credit: www. shutterstock.com, 177900254.)

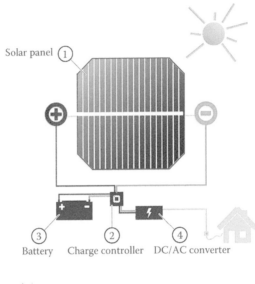

FIGURE 1.3   House equipped for the use of solar energy. (Credit: www.shutterstock.com, 186664391.)

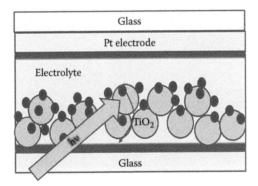

FIGURE 1.5   Basic structure of dye-sensitized solar cell.

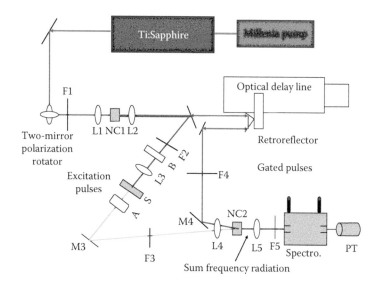

FIGURE 1.6 The basic setup for fluorescence time-resolved measurements.

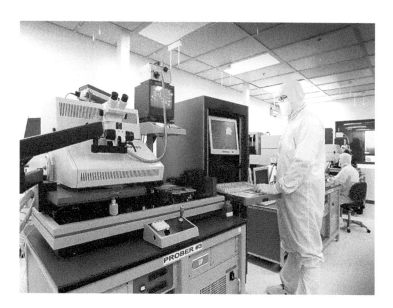

FIGURE 1.8 Silicon wafer fabrication and testing facility. (Credit: www.shutterstock.com, 19551091.)

FIGURE 1.9 High-performance computing has played a major role in the discovery of new organic solar materials. (Credit: www.shutterstock.com, 231835942.)

FIGURE 2.1 Fabrication of flexible solar cells.

FIGURE 3.1 Ancient sundial measuring device on a stone platform. (Credit: www.shutterstock.com, 21459691, Tarragona, Spain.)

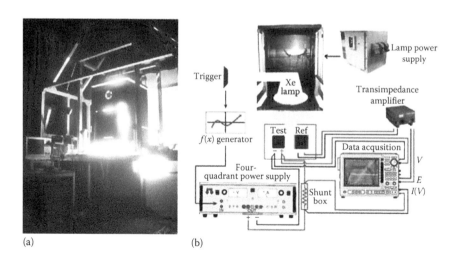

(a)   (b)

FIGURE 3.2 Solar simulators used to test OPV devices. (a) Solar simulator used to test PV efficiency. (b) Xe lamp apparatus to measure efficiency.

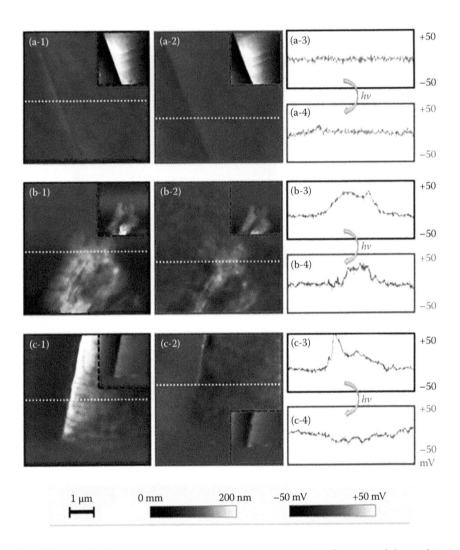

FIGURE 3.6 The Kelvin probe measurement of the polymer. Surface potential mappings in the dark (a-1, b-1, c-1) and under UV-B illumination (a-2, b-2, c-2) of three kinds of nano materials (a) SHTNTs, (b) N-TiO$_2$ NWs, and (c) N-TiO$_2$-P+NWs on gold thin film. The insets of surface potential images are topographic images (a-3, b-3, c-3) and (a-4, b-4, c-4) are surface potential values of the cross section obtained from the corresponding white-dotted line without and under UV-B illumination respectively. (From Wu, M.-C. et al., *J. Nanopart. Res.*, 16, 2143, 2014.)

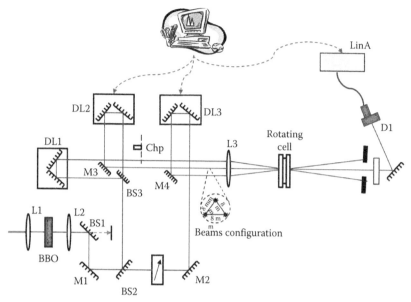

$\lambda_{\text{pump}} = \lambda_{\text{probe}} = 415$ nm, pump pulse energy ~0.5 nJ; $\Delta A/A \sim 10^{-7}$ S/N > 5

FIGURE 3.9 The optical apparatus for femtosecond transient absorption. (From Varnavski, O.P. et al., *J. Am. Chem. Soc.*, 124, 1736, 2002.)

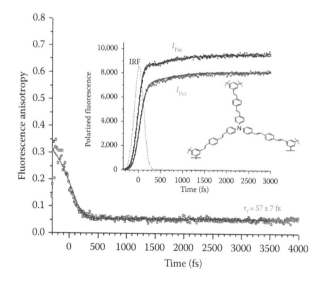

FIGURE 3.13 The fluorescence anisotropy of organic dendrimers for PV. (From Ransinghe, M. et al., *J. Phys. Chem.*, 108, 8543, 2004.)

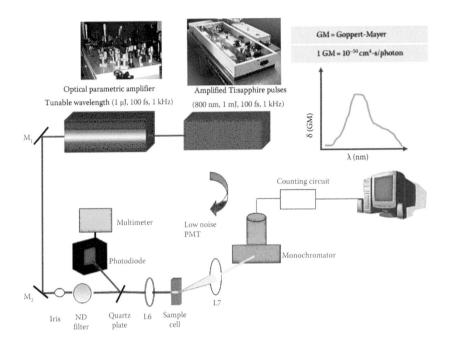

FIGURE 3.14 The two-photon excited fluorescence experimental apparatus.

FIGURE 4.1 Modern residential construction is now seriously considering solar energy in their design criteria. (Credit: www.shutterstock.com, 128612171.)

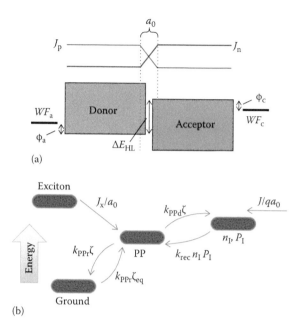

(a)

(b)

FIGURE 4.4 (a) Energy-level diagram showing the anode and cathode work functions. (b) Processes occurring within the HJ region. Excitons diffuse, with current density, J_X, to the HJ and undergo charge transfer to form polaron pairs. These may recombine at rate k_{PPr} or dissociate with rate k_{PPd}. (From Giebink, N.C. et al., *Phys. Rev. B*, 82, 15, 2010.)

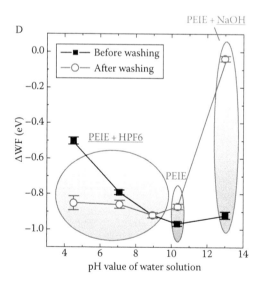

FIGURE 4.8 A new approach toward working electrodes for organic solar cells. (From Zhou, Y. et al., *Science*, 336(6079), 327, 2012.)

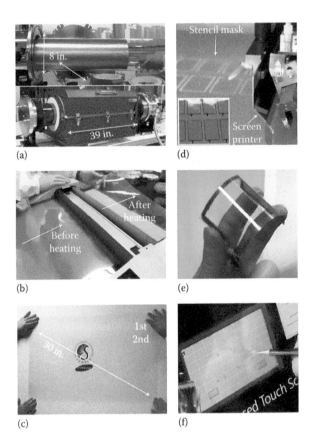

FIGURE 4.9 Roll-to-roll fabrication of substrate electrodes. (a) Copper foil wrapping around a 7.5 in. quartz tube to be inserted into an 8 in. quartz reactor. The lower image shows the stage in which the copper foil reacts with CH_4 and H_2 gases at high temperatures. (b) Roll-to-roll transfer of graphene films from a thermal release tape to a PET film at 120°C. (c) A transparent ultralarge-area graphene film transferred on a 35 in. PET sheet. (d) Screen printing process of silver paste electrodes on a graphene/PET film. The inset shows 3.1 in. graphene/PET panels patterned with silver electrodes before assembly. (e) An assembled graphene/PET touch panel showing outstanding flexibility. (f) A graphene-based touch screen panel connected to a computer with control software. (From Bae, S. et al., *Nat. Nanotechnol.*, 5, 574, 2010; Krebs, F.C. et al., *J. Mater. Chem.*, 19, 5442, 2009.)

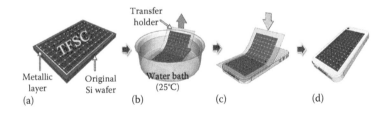

FIGURE 4.10 (a) As-fabricated TFSCs on the original Si/SiO$_2$ wafer. (b) The TFSCs are peeled off from the Si/SiO$_2$ wafer in a water bath at room temperature. (c) The peeled-off TFSCs are attached to a target substrate with adhesive agents. (d) The temporary transfer holder is removed, and only the TFSCs are left on the target substrate. (From Lee, C.H. et al., *Nature*, 2, 1000, 2012.)

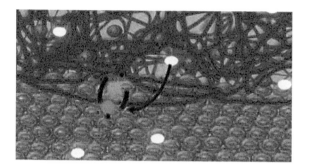

FIGURE 5.1 The interface of an organic solar device.

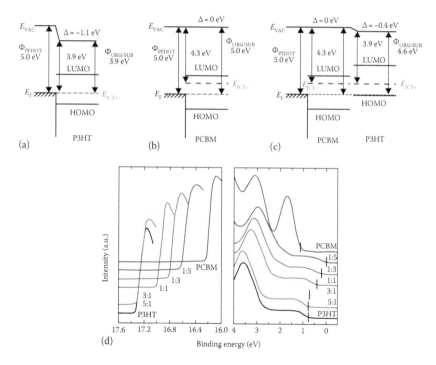

FIGURE 5.3 Schematic energy level alignment diagrams of (a) PEDOT:PSS/P3HT, (b) PEDOT:PSS/PCBM, (c) PEDOT:PSS/PCBM/P3HT as predicted by the ICT model, and (d) UPS spectra of P3HT, P3HT:PCBM, and PCBM films spin-coated on PEDOT:PSS. The PCBM concentration increases from bottom to top. (From Xu, Z. et al., *Appl. Phys. Lett.*, 95, 013301, 2009.)

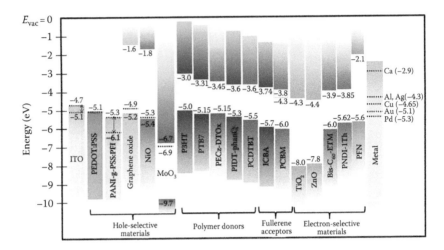

FIGURE 5.4 Schematic view of the energy gaps and energy levels of some of the components of recent OPVs including transparent electrodes, hole-selective materials, polymer donors, fullerene acceptors, electron-selective materials and metal electrodes. The dotted lines correspond to the work functions of the materials. (From Yip, H.L. and Jen, A.K.Y., *Energy and Environmental Science*, 5(3), 5994, 2012.)

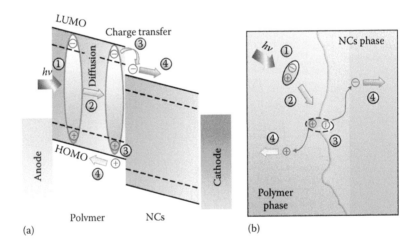

FIGURE 5.5 Schematic diagram of the photocurrent generation mechanism in bulk heterojunction hybrid solar cells. (a) Schematics of the different energy levels and the individual processes: exciton generation (1), exciton diffusion (2), charge transfer (3), charge carrier transport, and collection (4). (b) Schematic of a bulk heterojunction contact at the polymer–NC interface. The same processes mentioned in (a) are illustrated directly at the donor–acceptor interface. (From Zhou, Y.F. et al., *Energy Environ. Sci.*, 3(12), 1851, 2010.)

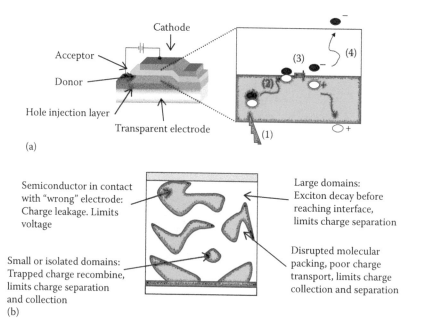

(a)

Cathode

Acceptor

Donor

Hole injection layer

Transparent electrode

(1) (2) (3) (4)

Semiconductor in contact
with "wrong" electrode:
Charge leakage. Limits
voltage

Large domains:
Exciton decay before
reaching interface,
limits charge separation

Small or isolated domains:
Trapped charge recombine,
limits charge separation
and collection

Disrupted molecular
packing, poor charge
transport, limits charge
collection and separation

(b)

FIGURE 5.6 (a) Schematic of a bilayer organic solar cell illustrating the main processes occurring in the photoactive layer: (1) absorption of photon to create an exciton, (2) exciton diffusion, (3) exciton splitting at the interface between donor and acceptor, and (4) diffusion and collection of charges. (b) Schematic of the side-view microstructure of a bulk heterojunction donor–acceptor blend film, illustrating the ways in which a non-optimum microstructure can affect device performance. (From Brabec, C.J. et al., *Chem. Soc. Rev.*, 40(3), 1185, 2011; Shaw, P.E. et al., *Adv. Mater.*, 20, 3516, 2008.)

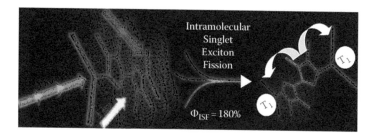

Intramolecular
Singlet
Exciton
Fission

T_1

T_1

$\Phi_{ISF} = 180\%$

FIGURE 6.1 Depiction of singlet exciton fission in an organic molecule.

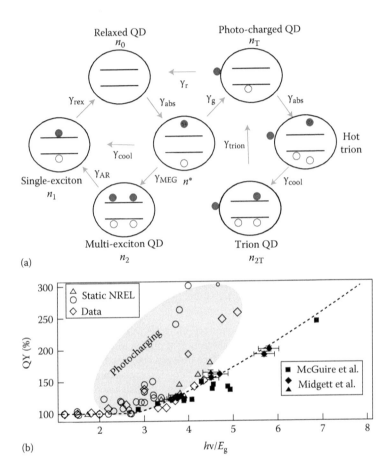

FIGURE 6.2 The mechanism and quantum yield dependence in semiconductor multi-exciton generation processes. (a) Relaxation pathways for excitons produced with high-energy photons. The photocharging pathway complicates the analysis of MEG yields and depends on the photon energy, photon fluence, and QD surface quality. The effects of photocharging on MEG yields can be reduced by flowing or stirring the samples during the experiment. (b) Photon-to-exciton QYs deduced from transient absorption data and results obtained on static solutions compared to flowing or stirring solutions.

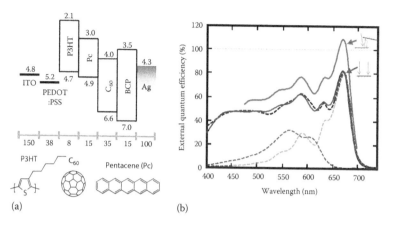

(a)

(b)

FIGURE 6.4 Organic solar cell schematic as well as external quantum efficiency with pentacene as the active layer. (a) Chemical structures and architecture of the solar cell with the thickness of each layer in nanometers and energy levels of the lowest unoccupied and highest occupied molecular orbitals in electron volts [12,18,20,29–31]. The anode is composed of indium tin oxide (ITO) and poly(3,4-ethylenedioxythiophene) poly(styrenesulfonate) (PEDOT:PSS). The cathode employs bathocuproine (BCP) and a silver cap. (b) External quantum efficiency of devices without optical trapping (blue line), and device measured with light incident at 10° from normal with an external mirror reflecting the residual pump light (red line). Optical fits from IQE modeling are shown with dashed lines: modeled pentacene EQE (blue dashes), modeled P3HT EQE (purple dashes), and modeled device EQE (black dashes) for comparison to the measured device efficiency without optical trapping.

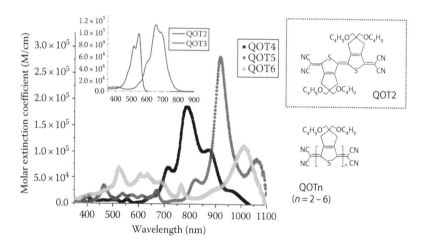

FIGURE 6.5 Organic QOT material and spectra for intramolecular singlet exciton fission. (From Varnavski, O. et al., *J. Phys. Chem. Lett.*, 6(8), 1375, 2015; Chien, A.D. et al., *J. Phys. Chem. C*, 119(51), 28258, 2015.)

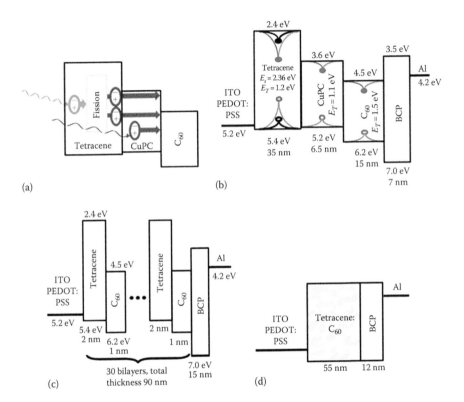

FIGURE 6.6 Device structures. (a) Schematic of a photovoltaic cell exhibiting singlet fission. Tetracene and CuPC are donor materials, and C_{60} is the acceptor. (b) Complete structure of the photovoltaic cell showing singlet and triplet exciton energies and lowest unoccupied and highest occupied molecular orbital energies. Singlets and triplets from tetracene diffuse through CuPC to the CuPC–C_{60} interface. BCP acts as an exciton and hole blocker. (c) Multijunction photodetector structure. (d) Bulk heterojunction solar cell with tetracene–C_{60} blend. (b: From Schaller, R.D. et al., *Nat. Phys.*, 1(3), 189, 2005; d: Chien, S.T. et al., *Laser Focus World*, 10(24), 2013.)

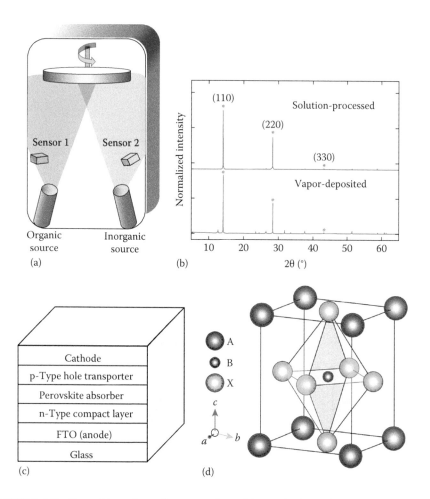

FIGURE 6.8 Perovskite solar cell construction with standard glass/metal substrates. (a) Dual-source thermal evaporation system for depositing the perovskite absorbers; the organic source was methylammonium iodide and the inorganic source PbCl₂. (b) X-ray diffraction spectra of a solution-processed perovskite film (blue) and vapour-deposited perovskite film (red). The baseline is offset for ease of comparison and the intensity has been normalized. (c) Generic structure of a planar heterojunction p–i–n perovskite solar cell. (d) Crystal structure of the perovskite absorber adopting the perovskite ABX3 form, where A is methylammonium, B is Pb and X is I or Cl. (From Liu, M.Z. et al., *Nature*, 501(7467), 395, 2013.)

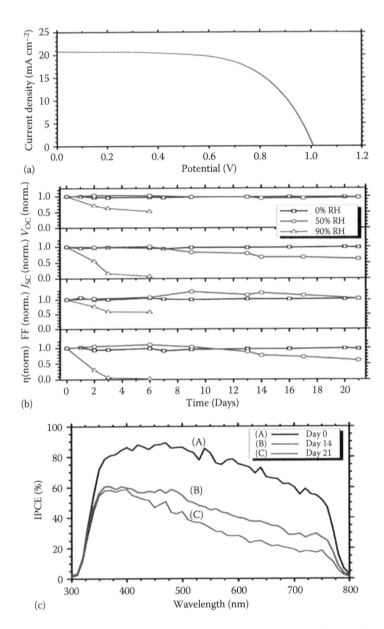

FIGURE 6.9 (a) Current–voltage characteristics of champion TiO₂/CH₃NH₃PbI₃/spiro-OMeTAD solar cell under 100 mW/cm² AM 1.5 irradiation. The scan was taken with a scan rate of 200 mV/s from forward bias to short-circuit conditions. The active area of the solar cell was masked and the area measured to be 0.113 cm². (b) Stacked plot of the normalized performance parameters with time for solar cells from Table 2 stored in 0%, 50%, and 90% RH. (c) IPCE spectra of perovskite solar cell that was stored in 50% RH taken following exposure to these conditions for (A) 0 days, (B) 14 days, and (C) 21 days.

FIGURE 7.1 Solar fuel at sunset. (Courtesy of T. Goodson III.)

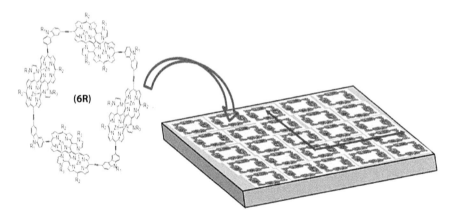

FIGURE 7.2 Square porphyrin structures as a new platform for new surfaces of solar materials. (Courtesy of T. Goodson III.)

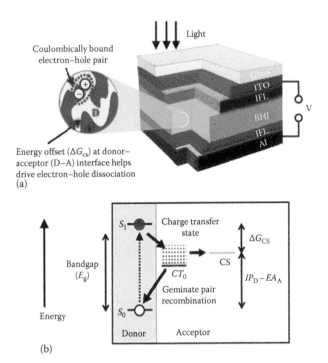

Light

Coulombically bound
electron–hole pair

Glass
ITO
IFL
BHJ
IFL
Al

V

Energy offset (ΔG_{cs}) at donor–
acceptor (D–A) interface helps
drive electron–hole dissociation
(a)

S_1

Charge transfer
state

ΔG_{CS}

Bandgap
(E_g)

CS

CT_0

$IP_D - EA_A$

Geminate pair
recombination

Energy

S_0

Donor   Acceptor

(b)

FIGURE 7.3 Charge transfer and interfacial effects in organic solar cells. Donor–acceptor energy offsets (ΔG_{CS}) and organic solar cell performance. (a) A bulk-heterojunction (BHJ) organic solar cell showing the Coulombically bound geminate electron–hole pair. Energy offsets at the interface of the donor (yellow) and acceptor (blue) materials drive electron–hole dissociation. ITO = indium tin oxide (a transparent electrode), IFL = interfacial layer, BHJ = bulk-heterojunction absorber layer, and Al = aluminum. (b) Energy landscape at the donor–acceptor interface in an organic solar cell. S_0 is the singlet ground state, S_1 is the singlet excited state, and CT_0 is the lowest energy singlet charge transfer state. The dotted red lines represent the multiple higher energy CT vibronic states. We define the energy offset ΔG_{CS} as $\Delta G_{CS} = E_{g,D} - (IP_D - EA_A)$, where IP_D is the ionization potential of the donor and EA_A is the electron affinity of the acceptor. ΔG_{CS} here represents the lower limit to the thermal energy available for dissociation. Note that triplet states are not shown. The donor and acceptor labels indicate the location of the electron. Note that "geminate pair dissociation" from a CT state to the charge separated state is the focus of the model here. (From Servaites, J.D. et al., *Energy Environ. Sci.*, 5(8), 8343, 2015.)

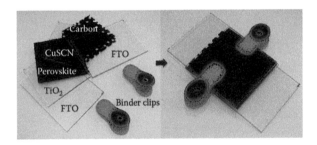

Carbon

CuSCN   FTO

Perovskite

TiO₂

FTO   Binder clips

FIGURE 7.4 Perovskite films and devices made for many scientists to appreciate. (From Patwardhan, S. et al., *J. Phys. Chem. Lett.*, 6(2), 251, 2015.)

several *p–n* junctions. Each junction is tuned to a different wavelength of light, reducing one of the largest inherent sources of losses, and thereby increasing efficiency. Traditional single-junction cells have a maximum theoretical efficiency of 34%, a theoretical "infinite-junction" cell would improve this to 87% under highly concentrated sunlight.[31–40]

In considering the current technology in organic and inorganic solar cells, it is agreed upon that the best lab examples of traditional silicon solar cells have efficiencies around 25%.[41] Although just recently, there have even been some smaller improvements in this regard for silicon, making the case for organic devices more urgent. Recent reports have shown that the crystalline silicon heterojunction structure adopted in photovoltaic modules commercialized as "Panasonic's HIT" have recently significantly reduced recombination loss, resulting in greater conversion efficiency.[42] The structure of an interdigitated back contact was adopted with the fabricated crystalline silicon heterojunction solar cells to reduce optical loss from a front grid electrode, a transparent conducting oxide layer, and a-Si:H layers as an approach for exceeding the conversion efficiency of 25%.[42] And the result was an improved short-circuit current and the world's highest efficiency of 25.6% for crystalline-silicon-based solar cells under 1-sun illumination.[42] This is quit an accomplishment. However one also notes that lab examples of multijunction cells have demonstrated performance over 43%.[43] Commercial examples of tandem cells (inorganic or silicon) are widely available at 30% under 1-sun illumination and improve to around 40% under concentrated sunlight.[43,44]

The use of concentrators has significantly progressed in the last 10 years as well. It has been suggested that the actual cost of photovoltaic power can be reduced with the use of organic solar concentrators. These are planar wave guides with a thin-film organic coating on the face and inorganic solar cells attached to the edges. Light is absorbed by the coating and reemitted into wave guide modes for collection by the solar cells. These devices may provide quantum efficiencies exceeding 50% and projected power conversion efficiencies as high as 6.8%.[45] These technologies were reported to enhance the power obtained by photovoltaic cells by a factor of 10.[45] However, this efficiency is gained at the cost of increased complexity and manufacturing price. To date, their higher price and lower price-to-performance ratio have limited their use to special roles, notably in aerospace where their high power-to-weight ratio is desirable. In terrestrial applications, these solar cells are used in operating plants all over the world. Tandem techniques can also be used to improve the performance of existing cell designs, although there are strict limits in the choice of materials. In particular, the technique can be applied to thin-film solar cells using amorphous silicon to produce a cell with about 10% efficiency that is lightweight and flexible. This approach has been used by several commercial vendors, but these products are currently limited to certain niche roles, like roofing materials. So this still leaves the important and difficult question as to what can be done to improve (and that is far reaching) the operation of multilayer systems which will have a large impact on a number of applications for solar power? One may look closer at bilayer devices with more physical and chemical intuition as to the choice of materials to use.

TANDEM ORGANIC CELL CONSTRUCTION

In general, for a typical bilayer device, a donor and an acceptor material are sand-wiched together with a planar interface. It is known that this is where the charge separation occurs, as well as a drop in the potential between the donor and acceptor.[46] The bilayer is sandwiched between two electrodes matching the donor HOMO and the acceptor LUMO, for an efficient extraction of the corresponding charge carri-ers. The bilayer device structure is schematically depicted in Figure 4.5, neglecting possible band bending due to energy level alignments.[47,48] As discussed earlier, for a typical *p/n*-junction formation, one requires the use of doped semiconductors with free charge carriers to form the electric field in the depleted region. In this case, the charge transfer in a bilayer heterojunction between undoped donor and acceptor materials is a result of the differences in the ionization potential and electron affinity of the adjacent materials.

In a typical experiment with a bilayer device with an organic material, the first process is a photon absorption in the donor D, which is followed by an electron being excited from the ground state to the first S_1 excited state $(S_0 \rightarrow S_1)$ of the molecule's electronic structure. Typically, electron and/or energy transfer takes place involving an acceptor molecule which is in close proximity to the donor.[49,50] Depending on the distance and electronic coupling between the donor and acceptor, the transfer will (or will not) take place efficiently. In general, this is still the same principle as the original ideas of Schottky. The electron and hole pair have a rather strong potential between them in the case of inorganic materials and are less tightly bound in the case of organic systems.[50]

The organic solar cells heavily rely on photo-generated excitons. As mentioned earlier, the length of travel of these excitons is rather short. It was found that the excitons closest to the interface were mainly involved. Scientists have tried to increase the surface area of the junction. With interdigitated structures, one can do this to a great extent. And for some organic devices there has been both electronic and spectroscopic data reported to suggest that the increase of the photovoltaic efficiency is related also to the existence of a weak charge transfer absorption state in the near-infrared region. The processes involving charge transfer are impor-tant in the donor–acceptor type bilayer devices, and many spectroscopic mea-surements (as referred to in previous chapters) have been used to find the rates

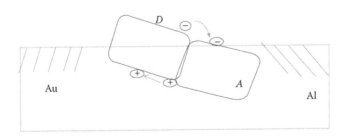

FIGURE 4.5　Bilayer heterojunction device construction.

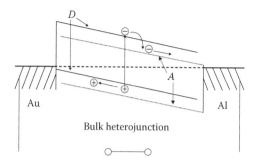

FIGURE 4.6 The bulk heterojunction construction.

of these transfer processes. One seeks to stabilize the charge-separated state by a repulsive interaction between the interface and the free charges. Therefore, the success of the D–A concept lies to a great extent in the relative stability of the charge-separated state. The recombination rate between holes in the donor and electrons in the acceptor is several orders of magnitude smaller than the forward charge transfer rate.

As introduced in previous chapters, one can also consider a bulk heterojunction architecture. In this case, one seeks to effectively mix the donor and acceptor components in a bulk volume so that each donor–acceptor interface is within a distance less than the exciton diffusion length of each absorbing site. In Figure 4.6, the situation is schematically shown for a bulk heterojunction device.[51,52] The bulk heterojunction device is similar to the bilayer device with respect to the donor–acceptor concept, but it exhibits an increased interfacial area where charge separation occurs. Due to the interface being dispersed throughout the bulk, no loss due to too small exciton diffusion lengths is expected. This is because all excitons should be dissociated within their lifetime. In this construction, the charges are also separated within the different phases, and hence recombination is reduced to a large extent. The mechanism of electron–hole recombination is different here as well. While it is still true that in the bilayer heterojunction, the donor and acceptor phase contact the anode and cathode selectively, the bulk heterojunction requires percolated pathways for the hole and electron.[51]

In the case of bulk heterojunctions, some researchers have modified the geometry of the device in order to "stack" the different layers for better performance. Reports have shown that high-efficiency organic photovoltaic cells are possible by stacking two hybrid planar-mixed molecular heterojunction cells in series.[53–55] For these devices, the absorption of incident light is maximized by locating the subcell tuned to absorb long-wavelength light nearest to the transparent anode and tuning the second subcell closest to the reflecting metal cathode to preferentially absorb short-wavelength solar energy. For example, in one structure the researchers used the donor copper phthalocyanine and the acceptor C_{60} to achieve a maximum power conversion efficiency of 5.7% under 1-sun simulated (AM1.5G solar illumination).[56] An open-circuit voltage of 1.2 V was obtained, doubling that of a single cell.[56] Also, it has been demonstrated that one can double the cell efficiency by stacking two thin cells in series with Ag nanoclusters between the subcells providing both optical field enhancement and efficient recombination sites for the photogenerated charges.[57] The photovoltage of this "tandem" cell was twice that of each individual cell (or subcell).[57] It was later shown that by stacking two CuPc/C_{60} hybrid cells in series with optimized optical absorption, an efficiency of 5.7% ± 0.3% under 1-sun simulated AM1.5G solar illumination is achieved, representing a 15% increase from a single hybrid PM-HJ cell.[58] Furthermore, the open-circuit voltage, V_{OC}, of the tandem cell is twice that of a single PV cell, reaching upwards of 1.2 V under high intensity illumination. These scientists showed that without including antireflection coatings on the substrates, organic PV cells with solar power conversion efficiencies exceeding 6.5% could be achieved with this geometry. This promising approach suggested that the power conversion efficiency of a tandem organic PV cell with hybrid heterojunctions has the potential for reaching that of a-Si cells, currently in production at 4%–6%. By applying antireflection coatings to the glass substrates, an additional 10% improvement to efficiencies is possible, suggesting that the tandem cell structure can attain efficiencies in excess of 7%.[59] However, the ultimate advantage of the asymmetric tandem cell structure lies in the ability to incorporate different D–A material combinations in the individual subcells to cover a broader solar spectral region than a typical CuPc/C_{60} cell system. It was found that a full solar spectral coverage may be optimally achieved by employing a three-subcell device, with two subcells absorbing across the blue to red, and a third subcell primarily absorbing in the near infrared. It has been recently illustrated that the process to make these cells can lead to high production yields and long operational lifetimes.[60] This is a particularly exciting avenue being pursued by researchers in the United States and elsewhere as asymmetric hybrid tandem cells have considerable potential for use in generating inexpensive, abundant electrical power from the clean and renewable energy generated by the sun (Figure 4.7).

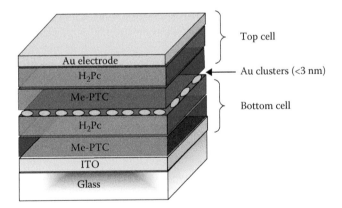

FIGURE 4.7 Geometry, optical density, and quantum efficiency of a tandem solar cell. (From Hiramoto, M. et al., *Chem. Lett.*, 20, 327, 1990.)

NEW ELECTRODES FOR ORGANIC SOLAR CELLS

One approach toward new methodologies to improve bulk heterojunctions involves the use of different electrodes. Indeed, organic-based thin-film optoelectronic devices, such as organic solar cells, organic light-emitting diodes, and organic thin-film transistors, have utilized considerable efforts to perfect this aspect of their devices given the great economic potential which may lead to a new generation of consumer electronic devices. While this point brings great enthusiasm, it is still the case that most optoelectronic devices require at least one electrode with a work function that is sufficiently low to either inject electrons into or collect electrons from the lowest unoccupied molecular orbital of a given organic semiconductor. Low-WF metals, such as alkaline earth metals (Ca, Mg) or metals co-deposited or coated with alkali elements (Li, Cs), meet this requirement.[61–63] One major issue with their use, however, is that they are chemically very reactive and easily oxidize in the presence of ambient oxygen and water.[64] Thus, their use in printed electronics presents limitations that can only be overcome by the fabrication of devices in an inert atmosphere and their subsequent encapsulation with barrier-coating technologies, which increases both the cost and complexity of the device architectures.

Several strategies for replacing low-WF metals have been explored. In one approach, a film of a conducting or semiconducting material, typically thicker than 10 nm and displaying a low WF, is coated on a high-WF electrode.[65] These films are typically made from an electron transport material, which mediates charge injection and transport between the higher-WF conductive electrode onto which it is coated and a semiconducting layer in the device. Their conductivity can be tuned from insulating to as far as conducting and their transparency can also be adjusted.[66] As they can be produced as n-type and p-type devices, they open a wide range of power-saving optoelectronic technological applications. Common examples of this approach have included coating indium tin oxide (ITO) with thin metal-oxide films—such as ZnO, In_2O_3, Al-doped ZnO, or In-doped ZnO.[66] Some have taken examples from the photoelectrochemical cell technology where novel surface preparations and modification of the semiconductor–liquid interface has proven to strongly govern surface-related processes in the cell.[67] In addition to this for the approach of direct hydrogen generation by photoelectrochemical water splitting, a customized tandem absorber structure to mimic natural photosynthesis has been reported.[68] The surface is modified chemically in this device and coupled with a catalyst deposition at ambient temperature yielding large photocurrents approaching the theoretical limit of the absorber and subsequently a solar-to-hydrogen efficiency of 14%.[68]

UNIVERSAL APPROACH TOWARD MODIFICATION OF ELECTRODES

As mentioned earlier, in more modern devices, what appears to be a "universal" approach to reducing the WF of a conductor has received great attention in the field of organic solar cells. In one reported case which has received surprisingly good results an ultrathin layer of a polymer containing simple aliphatic amine groups is adsorbed onto the conductor surface. This has focused new light on this matter

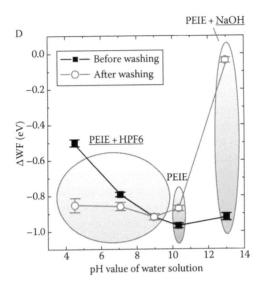

FIGURE 4.8 (See color insert.) A new approach toward working electrodes for organic solar cells. (From Zhou, Y. et al., *Science*, 336(6079), 327, 2012.)

with great enthusiasm. In contrast to the π-conjugated amine-containing small molecules and polymers considered earlier, the polymers exploited in this approach are large bandgap insulators and should not be regarded as charge-injection layers but rather as surface modifiers. The intrinsic molecular dipole moments associated with the neutral amine groups contained in such an insulating polymer layer, and the charge-transfer character of their interaction with the conductor surface, together reduce the work function of a wide range of conductors.[65] There are some commercially available polymer modifiers for this purpose, which have been suggested to be easily processed in air, from dilute solutions in environmentally friendly solvents. Scientists have found that the polymers containing simple aliphatic amine groups such as PEIE and PEI (Figure 4.8)[65] appear to be "universal" surface modifiers. The specific properties of the polymers can be further optimized for other applications, and conceptually, the approach could be applied to the development of polymers for high-WF electrodes. Their low cost and ease of processing make them compatible with roll-to-roll large-area mass production techniques and suited for organic or printed electronic devices (Figure 4.9).[69]

LOW-COST ITO-FREE ELECTRODE SOLAR SYSTEMS

As stated earlier, one of the essential requirements to be met for the production of cheap solar cells is the use of cost-effective materials. To fulfill this requirement, we have seen that some have taken the far-reaching approach of developing novel solar cell architectures in which the expensive transparent indium tin oxide electrode is replaced. One of the so-called ITO-free solar cell architectures developed by the Fraunhofer corporation is based on the substitution of the ITO electrode by a transparent polymer hole contact, which is supported by metal structures.[74] In order to

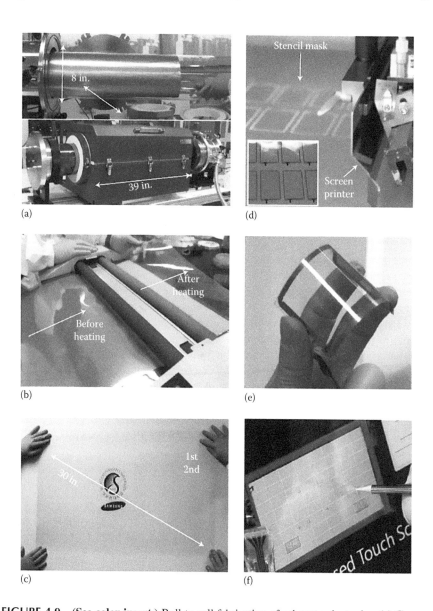

FIGURE 4.9 **(See color insert.)** Roll-to-roll fabrication of substrate electrodes. (a) Copper foil wrapping around a 7.5 in. quartz tube to be inserted into an 8 in. quartz reactor. The lower image shows the stage in which the copper foil reacts with CH_4 and H_2 gases at high temperatures. (b) Roll-to-roll transfer of graphene films from a thermal release tape to a PET film at 120°C. (c) A transparent ultralarge-area graphene film transferred on a 35 in. PET sheet. (d) Screen printing process of silver paste electrodes on a graphene/PET film. The inset shows 3.1 in. graphene/PET panels patterned with silver electrodes before assembly. (e) An assembled graphene/PET touch panel showing outstanding flexibility. (f) A graphene-based touch screen panel connected to a computer with control software. (From Bae, S. et al., *Nat. Nanotechnol.*, 5, 574, 2010; Krebs, F.C. et al., *J. Mater. Chem.*, 19, 5442, 2009.)

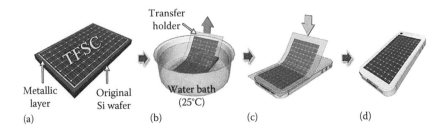

Transfer holder

Metallic layer

Original Si wafer

Water bath (25°C)

(a)   (b)   (c)   (d)

FIGURE 4.10 (See color insert.) (a) As-fabricated TFSCs on the original Si/SiO$_2$ wafer. (b) The TFSCs are peeled off from the Si/SiO$_2$ wafer in a water bath at room temperature. (c) The peeled-off TFSCs are attached to a target substrate with adhesive agents. (d) The temporary transfer holder is removed, and only the TFSCs are left on the target substrate. (From Lee, C.H. et al., *Nature*, 2, 1000, 2012.)

enable efficient production technologies, the deposition sequence of the electrodes was inverted in comparison to a standard organic solar cell. Based on this setup, Fraunhofer ISE holds a patent for a solar cell and module concept where the current is collected through small holes in the substrate and transported on the backside of the module, a so-called wrap-through contact (Figure 4.10). The first promising results with this concept have already been achieved at Fraunhofer ISE.[74] Another approach is to replace the ITO layer with a very thin silver film (<10 nm) embedded oxides.[75] This type of transparent contact can be adapted to function either as a hole or electron contact. Organic solar cells built with these electrodes can reach the same efficiency values compared to ITO-based devices.[75]

NEW GEOMETRIES FOR ORGANIC SOLAR CELL CONSTRUCTION

In connection with changes in the substrate electrodes, new ideas have appeared regarding the directional geometry of the solar organic device as well. Recent developments have celebrated new improvements of an inverted device that for laboratory devices is brought to the same level of performance as devices in a normal geometry. It has been shown that the limiting problems with operational stability for the devices with a normal geometry can be solved through stability studies using model atmospheres in an inverted device. Scientists and engineers have begun the task of transferring this process to a full roll-to-roll process entirely using solution processing on a flexible plastic substrate using slot-die coating and screen printing.[73] The initial performance for the roll-to-roll processed modules is acceptably close to the performance obtained for the laboratory cells in terms of both power conversion efficiency and operational stability (Figure 4.11).[73]

The use of an inverted geometry enables the use of such metals as silver as the hole-collecting electrode.[75] While the work function of silver is similar to that of aluminum and does not qualify as a metal with a high work function, the advantage of using silver as the metal electrode is that printable silver formulations are commercially available making the solution processing of the silver electrode a possibility. Researchers in this area have used silver grid systems as well. The use of the silver grid systems can serve several important purposes. In general, new results with

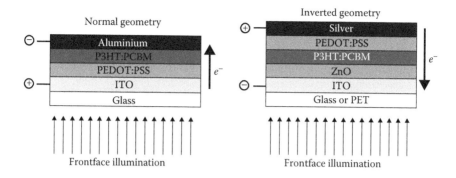

FIGURE 4.11 The diagrams for the normal and inverted devices. (From Krebs, F.C. et al., *J. Mater. Chem.*, 19, 5442, 2009.)

inverted devices have proven fruitful in increasing the efficiency.[76,77] Very talented chemists and engineers have developed new methodologies to test particular structure function relationships in these devices.[77]

As the reader can see, these new approaches to fabricate new organic solar cells have been productive. In some cases, the efficiency doubled due to the modification of a single-layer donor–acceptor device! The specific mechanisms of how some of these impressive modified devices work is still a matter of intense research. The simple single-layer device can be created and now be understood quite well (some 50 years later after the first single-layer device was reported). But for the processes of charge transfer, exciton diffusion, and polaron localization in the new multilayer cells, it may take more time to completely understand their physics. Indeed, the addition of many layers makes the optical and electronic spectroscopy substantially more complex. For example, the rates and cross sections of various processes of organic materials in single-layer devices are not to be assumed to be similar in a multilayer system. Generally, we can conclude that there is a major influence of donor and acceptor interactions on the morphology and subsequently the electronic processes in the organic multilayer cell devices. And we can further suggest that this connection between morphology and electronic interactions is still poorly understood. New molecular approaches to multilayer organic solar cells, while in its infancy, may provide an understanding of the macromolecular forces that drive domain separation and ultimately guide us to a new model to explain their electronic processes. Indeed, as stated at the beginning of this book, this is the time for new and far-reaching ideas. And perhaps looking at the interface more closely might be the first step in this direction.

REFERENCES

1. G. Williams, S. Sutty, and H. Aziz, Interplay between efficiency and device architecture for small molecule organic solar cells, *Physics Chemistry Chemical Physics*, 16, 11398, 2014.
2. H. Kallmann and M. Pope, Photovoltaic effect in organic crystals, *Journal of Chemical Physics*, 30, 585, 1959.

3. N. Geacintov, M. Pope, and H. Kallmann, Photogeneration of charge carriers in tetra-cene, *Journal of Chemical Physics*, 45, 2639, 1966.

4. B.A. Gregg, Excitonic solar cells, *Journal of Physical Chemistry B*, 107, 4688, 2003.

5. B.A. Gregg, M.A. Fox, and A.J. Bard, 2,3,7,8,12,13,17, 18-octakis (beta-droxyethyl) porphyrin (octaethanolporphyrin) and its liquid-crystalline derivatives: Synthesis and characterization, *Journal of American Chemical Society*, 111, 3024, 1989.

6. B.A. Gregg and M.C. Hanna, Comparing organic to inorganic photovoltaic cells: Theory, experiment, and simulation, *Journal of Applied Physics*, 93, 3605, 2003.

7. S. Ferrere, A. Zaban, and B.A. Gregg, Dye sensitization of nanocrystalline tin oxide by perylene derivatives, *Journal of Physical Chemistry B*, 101(23), 4490, 1997.

8. L.S. Hung, C.W. Tang, and M.G. Mason, Enhanced electron injection in organic elec-troluminescence devices using an Al/LiF electrode, *Applied Physics Letters*, 70(2), 152, 1997.

9. C.W. Tang and S.A. Vanslyke, Organic electroluminescent diodes, *Applied Physics Letter*, 51(12), 913, 1987.

10. J.D. Servaites, M. Ratner, and T.J. Marks, Organic solar cells: A new look at traditional models, *Energy and Environmental Science*, 4, 4410, 2011.

11. H. Wu et al., Low reflectivity and high flexibility of tin-doped indium oxide nanofiber transparent electrodes, *Journal of American Chemical Society*, 133(1), 27, 2011.

12. A. Kumar and C. Zho, The race to replace tin-doped indium oxide: Which material will win?, *ACS Nano*, 4(1), 11, 2010.

13. G.P. Crawford, *Flexible Flat Panel Displays*, Wiley Series in Display Technology, Wiley, Chichester, U.K., 2005.

14. Y. Sun and J.A. Rogers, Inorganic semiconductors for flexible electronics, *Advanced Materials*, 2007, 19, 1897.

15. S. Sz, *Physics of Semiconductor Devices*, John Wiley & Sons, New York, 1981.

16. H. Hoppe and N.S. Sariciftci, Organic solar cells: An overview, *Journal of Material Research*, 19(7), 1924, 2004.

17. D. Meissner, S. Siebentritt, and S. Gunster, Charge carrier pho-togeneration in organic solar cells, *Presented at the International Symposium on Optical Materials Technology for Energy Efficiency and Solar Energy Conversion XI: Photovoltaics, Photo-chemistry and Photoelectrochemistry*, Toulouse, France, 1992.

18. C.W. Tang and A.C. Albrecht, Photovoltaic effects of metal-chlorophyll-a-metal sand-wich cells, *Journal of Chemical Physics*, 62, 2139, 1975.

19. W. Schockley and H.J. Queisser, Detailed balance limit of efficiency of P-N junciton solar cells, *Journal of Applied Physics*, 32(3), 510, 1961.

20. P. Krogstrup, H.I. Jorgensen, M. Heiss, O. Demichel, J.V. Holm, M. Aagesen, J. Nygard, and A.F.I. Morral, Single-nanowire solar cells beyond the Shockley-Queisser limit, *Nature Photonics*, 7(4), 306, 2013.

21. A. Devos, Detailed balance limit of the efficiency of tandem solar cells, *Journal of Physical D, Applied Physics*, 13(5), 839, 1980.

22. L.C. Hirst and N.J. Ekins-Daukes, Fundamental losses in solar cells, *Progress in Photovoltaics*, 19(3), 286, 2011.

23. N. Grevesse and A.J. Sauval, Standard solar composition, *Space Science Reviews*, 85, 161, 1998.

24. N.C. Giebink, G.P. Wiederrecht, M.R. Wasielewski, and S.R. Forrest, Ideal diode equa-tion for organic heterojunctions. I. Derivation and application, *Physical Review B*, 82, 15, 2010.

25. T. Kirchartz, B.E. Pieters, J. Kirkpatrick, U. Rau, and J. Nelson, Recombination via tail states in polythiophene: Fullerene solar cells, *Physical Review B*, 83, 11, 2011.

26. C.W. Schlenker and M.E. Thompson, The molecular nature of photovoltage losses in organic solar cells, *Optics Communications*, 47(13), 3702, 2011.
27. E.L. Ratcliff, B. Zacher, and N.R. Armstrong, Selective inter layers and contacts in organic photovoltaic cells, *Journal of Physical Chemistry Letters*, 2(11), 1337, 2011.
28. C.W. Tang, 2-layer organic photovoltaic cell, *Applied Physics Letters*, 48(2), 183, 1986.
29. S.M. Menke and R.J. Holmes, Exciton diffusion in organic photovoltaic cells, *Energy and Environmental Science*, 7(2), 499, 2014.
30. H. Gommans, S. Schols, A. Kadashchuk, P. Heremans, and S.C.J. Meskers, Exciton diffusion length and lifetime in subphthalocyanine films, *Journal of Physics Chemistry C*, 113(7), 2974, 2009.
31. A. Hadipour, B. de Boer, and P.W.M. Blom, Organic tandem and multi-junction solar cells, *Advanced Functional Materials*, 18(2), 169, 2008.
32. J. Gilot, M.M. Wienk, and R.A.J. Janssen, Double and triple junction polymer solar cells processed from solution, *Applied Physics Letters*, 90, 14,2008.
33. K. Kawano, N. Ito, T. Nishimori, and J. Sakai, Open circuit voltage of stacked bulk heterojunction organic solar cells, *Applied Physics Letters*, 88, 7, 2006.
34. M. Lenes, L.J.A. Koster, V.D. Mihailetchi, and P.W.M. Blom, Thickness dependence of the efficiency of polymer: Fullerene bulk heterojunction solar cells, *Applied Physical Letters*, 88, 24,2006.
35. D.E. Markov, E. Amsterdam, P.W.M. Blom, A.B. Sieval, and J.C. Hummelen, Accurate measurement of the exciton diffusion length in a conjugated polymer using a heterostructure with a side-chain cross-linked fullerene layer, *Journal of Physical Chemistry A*, 109(24), 5266, 2005.
36. P. Peumans, V. Bulovic, and S.R. Forrest, Efficient photon harvesting at high optical intensities in ultrathin organic double-heterostructure photovoltaic diodes, *Applied Physics Letters*, 76, 2650, 2000.
37. S.E. Shaheen, R. Radspinner, N. Peyghambarian, and G.E. Jabbour, Fabrication of bulk heterojunction plastic solar cells by screen printing, *Applied Physics Letters*, 79(18), 2996, 2001.
38. J.G. Xue, B.P. Rand, S. Uchida, and S.R. Forrest, A hybrid planar-mixed molecular heterojunction photovoltaic cell, *Advanced Materials*, 17(1), 66, 2004.
39. G. Yu, J. Gao, J.C. Hummelen, F. Wudl, and A.J. Heeger, Polymer photovoltaic cells: Enhanced efficiencies via a network of internal donor-accepor heterojunctions, *Science*, 270(5243), 1789, 1995.
40. W.J.E. Beek, M.M. Wienk, M. Kemerink, X.N. Yang, and R.A.J. Janssen, Hybrid zinc oxide conjugated polymer bulk heterojunction solar cells, *Journal of Physical Chemistry B*, 109(19), 9505, 2005.
41. A. Rohatgi and P. Raichoudhury, High efficiency silicon solar cells-development, current issues and future directions, *Solar Cells*, 17(1), 119, 1986.
42. K. Masuko et al., Achievement of more than 25% conversion efficiency with crystalline silicon heterojunction solar cell, *IEEE Journal of Photovoltaics*, 4(6), 1433, 2014.
43. M. Treacy, Sharp hits solar cell efficiency record of 43.5%, Treebugger.com, May 31, 2012.
44. R.R. King, D.C. Law, K.M. Edmondson, C.M. Fetzer, G.S. Kinsey, H. Yoon, R.A. Sherif, and N.H. Karam, 40% efficient metamorphic GaInP/GaInAs/Ge multijunction solar cells, *Applied Physics Letters*, 90, 18, 2007.
45. M.J. Currie, J.K. Mapel, T.D. Heidel, S. Goffri, and M.A. Baldo, High-efficiency organic solar concentrators for photovoltaics, *Science*, 321(5886), 226, 2008.
46. S. Sun and N.S. Sariciftci, *Organic Photovoltaics: Mechanisms, Materials, and Devices*, CRC Press, Boca Raton, FL, March 29, 2005.

47. A. Zaban, S.G. Chen, S. Chappel, and B.A. Gregg, Bilayer nanoporous electrodes for dye sensitized solar cells, *Chemical Communications*, 22, 2231, 2000.
48. A.L. Ayzner, C.J. Tassone, S.H. Tolbert, and B.J. Schwartz, Reappraising the need for bulk heterojunctions in polymer-fullerene photovoltaics: The role of carrier transport in all-solution-processed P3HT/PCBM bilayer solar cells, *Journal of Physical Chemistry C*, 113(46), 20050, 2009.
49. K.H. Lee et al., Morphology of all-solution-processed "Bilayer" organic solar cells, *Advanced Materials*, 23(6), 766, 2011.
50. B.G. Kim, C.G. Zhen, E.J. Jeong, J. Kieffer, and J. Kim, Organic dye design tools for efficient photocurrent generation in dye-sensitized solar cells: Exciton binding energy and electron acceptors, *Advanced Functional Materials*, 22(8), 1606, 2012.
51. S.H. Park, A. Roy, S. Beaupre, S. Cho, N. Coates, J.S. Moon, D. Moses, M. Leclerc, K. Lee, and A.J. Heeger, Bulk heterojunction solar cells with internal quantum efficiency approaching 100%, *Nature Photonics*, 3(5), 297–305, 2009.
52. M.C. Scharber, D. Wuhlbacher, M. Koppe, P. Denk, C. Waldauf, A.J. Heeger, and C.L. Brabec, Design rules for donors in bulk-heterojunction solar cells: Towards 10% energy-conversion efficiency, *Advanced Materials*, 6, 789, 2006.
53. Y.F. Li, Molecular design of photovoltaic materials for polymer solar cells: Toward suitable electronic energy levels and broad absorption, *Accounts of Chemical Research*, 45(5), 723, 2012.
54. T. Ameri, N. Li, and C.J. Brabec, Highly efficient organic tandem solar cells: A follow up review, *Energy and Environmental Science*, 6(8), 2390, 2013.
55. Y. Yamamoto, T. Fukushima, A. Saeki, S. Seki, S. Tagawa, N. Ishii, and T. Aida, Molecular engineering of coaxial donor-acceptor heterojunction by coassembly of two different hexabenzocoronenes: Graphitic nanotubes with enhanced photoconducting properties, *Journal of the American Chemical Society*, 129(30), 9276, 2007.
56. P. Sullivan, S. Heutz, S.M. Schultes, and T.S. Jones, Influence of codeposition on the performance of CuPc–C60 heterojunction photovoltaic devices, *Applied Physics Letters*, 84(7), 1210, 2004.
57. A. Yakimov and S.R. Forrest, High photovoltage multiple-heterojunction organic solar cells incorporating interfacial metallic nanoclusters, *Applied Physical Letters*, 80, 1667, 2002.
58. P. Kumar and S. Chand, Recent progress and future aspects of organic solar cells, *Progress in Photovoltaics: Research Applied*, 20, 377–415, 2012.
59. M. Agrawal, P. Peumans, Broadband optical absorption enhancement through coherent light trapping in thin-film photovoltaic cells, *Optics Express*, 16(8), 5385, 2008.
60. A. Goetzberger, C. Hebling, H.W. Schock, Photovoltaic materials, history, status and outlook, *Materials Science and Engineering Reports*, 40(1), 1–46.
61. K.M. Coakley and M.D. McGehee, Conjugated polymer photovoltaic cells, *Chemistry of Materials*, 16(23), 4533, 2004.
62. W.J. Potscavage, A. Sharma, and B. Kippelen. Critical interfaces in organic solar cells and their influence on the open-circuit voltage, *Accounts of Chemical Research*, 42(11), 1758, 2009.
63. H. Park, P.R. Brown, V. Buloyic, and J. Kong, Graphene as transparent conducting electrodes in organic photovoltaics: Studies in graphene morphology, hole transporting layers, and counter electrodes, *Nano Letters*, 12(1), 133, 2012.
64. N.R. Armstrong, P.A. Veneman, E. Ratcliff, D. Placencia, M. Brumbach, Oxide contacts in organic photovoltaics: Characterization and control of near-surface composition in indium-tin oxide (ITO) electrodes, *Accounts of Chemical Research*, 42(11), 1748, 2009.
65. Y. Zhou et al., A universal method to produce low–work function electrodes for organic electronics, *Science*, 336(6079), 327–3323.

66. A. Stadler, Transparent conducting oxides—An up-to-date overview, *Materials*, 5, 661–683, 2012.
67. G. Néstor, M.S. Prévot, and K. Sivula, Surface modification of semiconductor photo-electrodes, *Physical Chemistry Chemical Physics*, 17, 15655, 2015.
68. M.M. May, H.-J. Lewerenz, D. Lackner, F. Dimroth, and T. Hannappel, Efficient direct solar-to-hydrogen conversion by in situ interface transformation of a tandem structure, *Nature Communication*, 6, 8286, 2015.
69. F.C. Krebs, S.A. Gevorgyan, and J. Alstrup, A roll-to-roll process to flexible polymer solar cells: Model studies, manufacture and operational stability studies, *Journal of Materials Chemistry*, 19(30), 5442, 2009.
70. H. Park, S. Chang, M. Smith, S. Gradecak, and J. Kong, Interface engineering of graphene for universal applications as both anode and cathode in organic photovoltaics, *Scientific Reports*, 3, 1581, 2013.
71. M. Hiramoto, M. Suezaki, and M. Yokoyama, Effect of thin gold interstitial-layer on the photovoltaic properties of tandem organic solar cell, *Chemical Letters*, 20, 327, 1990.
72. S. Bae et al., Roll-to-roll production of 30-inch graphene films for transparent electrodes, *Nature Nanotechnology*, 5, 574–578, 2010.
73. F.C. Krebs, S.A. Gevorgyana, and J. Alstrupa, A roll-to-roll process to flexible polymer solar cells: Model studies, manufacture and operational stability studies, *Journal of Material Chemistry*, 19, 5442–5451, 2009.
74. Fraunhofer Inc., The Fraunhofer ITO free polymeric electrode substrate, http://www.academy.fraunhofer.de/en/energy_sustainability/photovoltaic.html. (Accessed June 14, 2016.)
75. D. Lee, D.-Y. Youn, Z. Luo, and I.-D. Kim, Highly flexible transparent electrodes using a silver nanowires-embedded colorless polyimide film via chemical modification, *RSC Advances*, 6, 30331, 2016.
76. J. Subbiah, C.M. Amb, I. Irfan, Y. Gao, R.J. Reynolds, and F. So, High-efficiency inverted polymer solar cells with double interlayer, *ACS Applied Materials Interfaces*, 4(2), 866–870, 2012.
77. C.E. Small, S. Chen, J. Subbiah, C.M. Amb, S.-W. Tsang, T.-H. Lai, J.R. Reynolds, and F. So, High-efficiency inverted dithienogermole–thienopyrrolodione-based polymer solar cells, *Nature Photonics*, 6, 115–120, 2012.
78. C.H. Lee, D.R. Kim, I.S. Cho, N. William, Q. Wang, and X. Zheng, Peel-and-stick: Fabricating thin film solar cell on universal substrates, *Nature*, 2, 1000, 2012, DOI: 10.1038/srep01000.

5 Interface Engineering in Organic Solar Cells

As mentioned in the previous chapter, the process of charge transfer at an interface is key to the understanding of the function of organic solar cells. While the process is well understood for inorganic interfaces,[1] such as in the case of crystalline inorganic semiconductors,[2] the situation is less clear for interfaces that involve organic materials. Interface engineering, as it is now called, can provide scientists and engineers with a constructive method to facilitate carrier extraction in order to enhance the organic solar cell efficiency. Several varieties of interface materials have been developed in recent years to systematically enhance the efficiency.[3,4] Over this time of development, it has been found that these interface materials may be both conducting and nonconducting as well as semiconducting. As the reader will see in this chapter, depending on the chemical structure and properties of the interfacial material, one can strongly affect the polarity of the solar cell device, open-cell voltage, and the contact and thickness properties at the interface. As mentioned in previous chapters, if one desires a clear physical mechanism of the charge carrier dynamics at the interface, the choice of measurement techniques is very important. The reader will also see new techniques to probe the interface have provided new insights into the structure–function relationships important to efficient charge transfer across the interface. Also, it was mentioned in the previous chapter that some bulk heterojunction (BHJ) photovoltaic cells with efficiencies approaching 10% have been observed with interface engineering. Indeed, if it is possible to tailor BHJs with tandem cells, then the efficiency might also be substantially increased with careful interfacial engineering. In considering this possibility, it is now clear that the interface structure and function will be very important in this process. As the reader will see, this might push organic solar cells into efficiencies comparable to the best cells commercially sold presently. Finally, the reader will also appreciate that in thinking about interface materials it is not only important to consider the possible increase in efficiency, but also it is important to consider the effect of the interface material on the stability and lifetime of the organic solar cell device (Figure 5.1).

BASIC MECHANISMS INVOLVING INTERFACE LAYERS IN ORGANIC SOLAR CELLS

In order to fully understand how one might utilize interface materials for improved solar cell devices, it is important to first consider a few basic principles regarding the processes of charge carriers at the interface. One can consider in most devices two kinds of interactions, organic to metal and organic to organic interactions.[5] With these interactions in mind, one can consider that at the interface there is a dipole created, which has a specific polarization. Several important factors that have been investigated

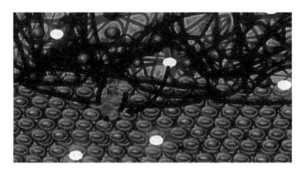

FIGURE 5.1   **(See color insert.)** The interface of an organic solar device.

with organic systems play a role at the interface polarization. There have been reports of factors relating to the formation of the interface state, alignment of the permanent dipole of the organic material, photoinduced charge transfer across the interface, perturbations of the electron cloud of the metal surface by the organic material, and, finally, the ever possible chemical reactions at the interface.[6–10] The alignment is typically explained by a vacuum level alignment model.[11] But due to more recent XPS studies, this model has been considered rather limited in describing the process of the formation of dipoles at the interface between the metal and organic or the organic and bottom electrode in the case of the use of indium tin oxide (ITO).[12] The formation of interface changes in potential ultimately lead to small changes in the electronic structure highest occupied molecular orbital-lowest unoccupied molecular orbital (HOMO-LUMO levels) of the organic layer. The process of photoinduced charge transfer happens near the interface. Due to the relatively low dielectric constant of the donor–acceptor system, the electron and hole pair can still be bound to each other even after the photoinduced charge transfer process occurs. However, many systems show a large efficiency for charge separation.[13] The issue of thermodynamics plays a major role here as well. The organic material and metal contact must be in thermal equilibrium at the interface. As we will see in this chapter, with procedures such as annealing and other techniques, important processes in the dynamics of the formation of the interface can be altered. However, it is the final state in a useful material which must be in thermal equilibrium for the interface dipole to be formed. Models have been suggested to explain the thermal process of the formation of interface dipoles.[14] In particular, if one considers the charge transfer and thermodynamics equilibrium between the metal and organic material, one can find a distance dependence on the interface dipole given as[15]

$$V_{\text{dipole}} = \left(\frac{-ed_{\text{MS}}M_{\text{b}}}{\epsilon_{\text{it}}} \right) \left(\frac{\varphi_{\text{M}} - \left(IP - E_g / 2 \right)}{E_g + k} \right) \tag{5.1}$$

Here, d_{MS} is the distance between the metal and the semiconductor material.[15] The k factors in Equation 5.1 are related to the hopping dynamics in the semiconductor and in the metal. Charge transfer across the interface is distance dependent and governed by the ionization potential and the work function of the metal. Also note that the for a dipolar interface, the half gap energy ($E_g/2$) is important.[15] There has been great discussion regarding excess energy. Generally, one excites above the bandgap in the device. Thus, excess energy discussions in organic structures have focused on the storage of vibrational energy (or vibronic contributions). This has led to the suggestion that there may be a structure–function relationship to those molecules which can store and use this excess energy more efficiently.[16]

Researchers have suggested the use of electrostatic models which have the ability to predict the formation of the charge transfer states at the interface with localized carriers.[17] It is not surprising that the charge transfer state persists at the organic–organic interfaces. Several reports have demonstrated that for the dynamics of exciton dissociation and charge separation in many organic semiconductor donor–acceptor systems, the formation of spatially separated carriers and bound charge transfer pairs strongly depends on the ordering of both the acceptor and donor as well as the order at their interface.[18–20] The degree of delocalized charge transfer states plays a central role for free carrier generation. As mentioned previously, the excess energy aids in the charge separation process for typical donor–acceptor systems. In the design of organic systems chapter we learned that it is possible to design donor–acceptor systems with strong delocalized charge transfer character. Indeed, with the proper building blocks, one can construct molecular topologies for enhanced delocalized charge transfer states in oligomers, dendrimers, and polymers.[21–24] The ability to create these systems in order to engineer the interface is a major goal in the field of organic solar cells.

BUILDING DELOCALIZED CHARGE TRANSFER STATES AT THE INTERFACE

At several points in this exploration of organic solar cell physics in this book, the role of delocalized states on the charge transfer process in an organic solar cell has come into focus. It can have an effect on the efficiency of the cell. The charge transfer state is, in most simple terms, a pair of opposite charges. The polarization energy and bandwidth determine the localization properties of the charge transfer state.[25] One can estimate the charge transfer polarization energy by adding together the hole polarization energy and the electron polarization energy.[25] In the case of inorganic semiconductors, one usually considers that carriers move in a band or by hopping, depending on the energy gain of the system by either polaron formation or by carrier delocalization.[26] Stabilization energy due to delocalization in a band is given by $E/2$, where E is the bandwidth. As mentioned earlier, if the magnitude of polarization is less than this value ($<E/2$), then it is found that the carrier becomes delocalized. For molecules such as pentacene, fullerenes, and phthalocyanines mentioned previously, the hole polaron binding energy is found to be lower than that for molecules with many internal degrees of freedom.[27–30] It has been postulated that if a low polaron binding energy is essential for delocalization, then the rigid molecules are more likely to form delocalized charge transfer states than their nonrigid counterparts. It is also suggested that the bandwidth is larger in rigid molecules as well. It has been recently found that although the delocalized states are extremely

short-lived (<1 ps), they enable charges to override the otherwise dominant Coulomb interaction.[31] Thus, it has been found that those materials that support delocalized charge wave functions and have low reorganization energies due to structural rigidity and suppressed torsion relaxation should be targeted for the next generation of organic solar devices.[32] This approach would mitigate the problem of polaron formation and allow for efficient charge separation with minimal band offsets, greatly increasing the open-circuit voltage and efficiency of organic solar cells.[32] Rigid molecules also lead to other advantageous characteristics such as smaller reorganization energies, which translate into higher carrier mobility.[33] In this context, the issue at the interface has been considered both experimentally and theoretically. For example, new studies have appeared to investigate the role of charge delocalization on the nature of the charge-transfer states in the case of model (Figure 5.2) complexes consisting of several pentacene molecules and one fullerene (C_{60}) molecule, which are representative of donor/acceptor heterojunctions.[34] The energies of the charge transfer states were examined by means of the time-dependent density functional theory (TD-DFT) using the long-range-corrected functional. TD-DFT is a quantum mechanical theory often used in physics and chemistry to investigate the properties and dynamics of many-body systems. This method can provide information about the electronic structure in the presence of time-dependent potentials such as electric or magnetic fields. The effect of the electric field on molecules can be studied with TD-DFT to gain information regarding the excitation energies, frequency-dependent response properties, and absorption spectra. In the pentacene study, a general description of how the nature of the charge transfer states is affected by molecular packing in terms of the interfacial donor/acceptor orientations was examined.[34] What was found from this investigation was that the delocalization effects as a function of system size are more pronounced in the edge-on than in the face-on molecular configurations. Also, an interesting result suggested that the decrease in donor−acceptor interactions due to an increase in the distance between donor and acceptor can lead to the enhancement of charge delocalization.[34] This was

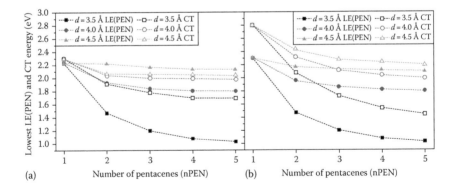

FIGURE 5.2 Evolution of the energies of the lowest LE(nPEN) and CT states in (a) nPEN−C60/P5 and (b) nPEN−C60/T5 with variation in the distance between adjacent pentacenes $d = 3.5$, 4.0, and 4.5 Å. (From Yang, B. et al., *J. Phys. Chem. C*, 118(48), 27648, 2014.)

not expected. However, the main impact on the energy of charge transfer states is due to the change in electrostatic interactions. The increase in the energy of charge transfer states parallels, in this case, the decrease in electron−hole binding energy. For example, the enhancement in hole delocalization because of an increase in the size of the pentacene cluster or because of an increase of inter-pentacene molecular interactions results in a lowering of the energy of the lowest charge transfer state accompanied by a decrease in electron−hole binding energy.[34] More experiments and theory are necessary to investigate this critical issue. Newer experimental techniques which take advantage of some of the tools mentioned in Chapter 3 are presently being modified for the purpose of studying interfacial charge transfer process. Hopefully, it is in this area that there may be future contributions toward the mechanisms and later possible control of the dynamics of charges at the interface in organic solar cell devices. For a good interface design, the fundamental interfacial atomic and electronic structures have to be fully understood and controlled. Interfacial interactions and the lattice match are always limiting factors in creating a nearly perfect device. And interface engineering can be an effective approach to enhance the stability of the cell.[35]

ENERGY LEVEL TUNING AT THE ELECTRODE/ACTIVE LAYER INTERFACE AND THE INTEGER CHARGE TRANSFER MODEL

As the reader is aware by now, the understanding of the interfacial electronic structure and energy level alignment at the electrode–BHJ contact is important for designing better contacts to improve charge extraction for organic solar devices. However, the nature of electronic contact formation between the electrode and active layer can be rather complex. As mentioned earlier, at the interface there are a number of important interfacial effects including charge transfer, dipole formation, and the formation of interface states, which can occur depending on the type and strength of interactions between the two materials and the order of contact formation such as organic-to-metal or metal-to-organic. As mentioned in the model description of the interface dipole, one of the most widely considered interfacial effects for organic devices is the Fermi-level pinning of electrodes to the negative integer charge transfer state of the acceptor and positive integer charge-transfer (ICT) state of the donor due to the corresponding electron and hole transfer from the electrode to the organic layer.[36,37] It has been suggested that the integer charge-transfer model is particularly suitable to describe interfaces formed by the solution processing of materials.[37] Recently, researches have attempted to classify the fundamental limiting cases at which the interface charge transfer model might predict the organic–substrate interface.[38–40] The first case considers the situation when the substrate work function is lower than the energy of the charge transfer state (E_{ICT}) of the organic semiconductor.[38] Here, the electrons will transfer from the substrate to the organic semiconductor and the Fermi level will eventually be pinned to the E_{ICT}. Another situation of the organic/substrate may be in the case when the substrate work function is in between E_{ICT} and the positive integer charge transfer state (E_{ICT+}) and no Fermi-level pinning occurs.[39] Finally, the situation when the substrate work function is higher than the E_{ICT+} of the organic semiconductor has also been considered. In this case, the holes will transfer from the substrate to the organic semiconductor and the Fermi level will eventually be pinned to the E_{ICT+}.[39] Both experiment and theory have verified these limiting cases. For example, photoelectron spectroscopy was used to map out the energy level alignment of conjugated polymers at various organic–organic and hybrid interfaces.[41–43] Researchers investigated the hole–injection interface between the substrate and a light-emitting polymer. What was found was that two different alignment regimes were observed. In particular, the researchers found a vacuum-level alignment, which corresponds to the lack of vacuum-level offsets Schottky–Mott limit.[41] Fermi-level pinning was observed where the substrate Fermi level and the positive polaronic level of the polymer align. These observations were rationalized in terms of spontaneous charge transfer whenever the substrate Fermi level exceeded the positive polaron/bipolaron formation energy per particle.[41] In a similar report, ultraviolet photoelectron spectroscopy measurements in combination with the ICT model was used to obtain the energy-level alignment diagrams for two common types of bulk heterojunction solar cell devices based on the well-known poly(3-hexylthiophene) (P3HT) or poly(2-methoxy-5-(3,7-dimethyloctyloxy)-1,4-phenylene vinylene) (MDMO-PPV) as the donor polymer and (6,6)-phenyl-C61-butyric acid methyl ester as the acceptor molecule (Figure 5.3).[42] This report found values for the integer charge

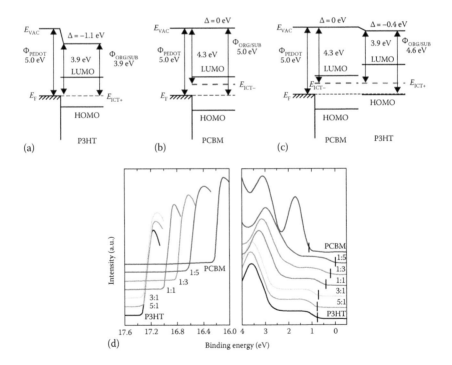

FIGURE 5.3 **(See color insert.)** Schematic energy level alignment diagrams of (a) PEDOT:PSS/P3HT, (b) PEDOT:PSS/PCBM, (c) PEDOT:PSS/PCBM/P3HT as predicted by the ICT model, and (d) UPS spectra of P3HT, P3HT:PCBM, and PCBM films spin-coated on PEDOT:PSS. The PCBM concentration increases from bottom to top. (From Xu, Z. et al., *Appl. Phys. Lett.*, 95, 013301, 2009.)

transfer states for specific organic materials used in many bulk heterojunction solar cells. These values were further applied to the ICT model to derive an energy-level alignment diagram for both MDMO-PPV:PCBM- and P3HT:PCBM-based bulk heterojunction solar cells.[43] What was found in both cases was that there are ground-state interface dipoles at the organic–organic heterojunctions that can help improve exciton dissociation and charge separation in the devices.[41] It was speculated that the P3HT:PCBM combination is particularly suitable, as its interface dipole involves the population of the most tightly bound CT electron–hole pairs that can be created at the interface. This results in the creation of hotter (less tightly bound) CT states upon exciton dissociation.[42]

While the ICT model has been successful in predicting the dominant charge mobility mechanisms in many bulk heterojunction solar cells, there are other factors which may affect the interaction at the interface and thus the efficiency. For example, some reports have looked at the importance of film morphology on the interface effects in polymer blends.[43–46] It was found that the control of blend morphology at the microscopic scale is critical for optimizing the power conversion efficiency of polymer solar cells.[44] In the case of bulk heterojunctions of region-regular P3HT and a soluble fullerene derivative ([6,6]-phenyl-C_{61}-butyric acid methyl ester, PCBM), both blend morphology and photovoltaic device performance are influenced by various treatments. These treatments include the choice of solvent, rate of drying, thermal annealing, and vapor annealing.[44,47] It was demonstrated that using a set of techniques may lead to a common arrangement of the components, which consists of a vertically and laterally phase-separated blend of crystalline P3HT and PCBM.[44,47] Many scientists have concluded that the morphology consists of an initial crystallization of P3HT chains followed by the diffusion of PCBM molecules to nucleation sites at which aggregates of PCBM then grow.[47] Indeed all of these configurations result in a different organic interface/metal interaction and will affect the use of the ICT model dramatically.

The role of morphology, long-range order, and delocalization are all intimately connected to the properties of the interface and are system dependent. In the area of material design and preparation, the focus should be on preparing materials with as much long-range order as possible and with extended conjugation while preserving a large interface area.[45] In addition to the high carrier mobility, the quality of the donor/acceptor interface also plays an important role. The development of interfaces with fewer defects is a necessary requirement. Generally, mutually compatible (mixable) materials provide interfaces with fewer defects.[13,48] However, a balance between defect-free interfaces and large interfacial area is required for efficient solar cells.[48,49]

MOLECULAR DESIGN OF INTERFACE MATERIALS

As one might expect, the experimental investigation of the effect of interfacial interactions on the efficiency of a solar cell is rather tedious. To understand how interlayers affect the organic solar device performance, one needs to evaluate the short-circuit current (J_{SC}), open-circuit voltage (V_{OC}), and fill factor very carefully. As discussed earlier, the V_{OC} is determined by the level alignment between the organic donor materials used in the device and typically a fullerene (C_{60}). The J_{SC} is determined by the light harvesting efficiency and the charge separation efficiency under a high extraction field near the short-circuit condition. The fill factor (FF) is determined by the device series resistance, dark current, and the charge recombination rate and extraction efficiency at low internal fields close to the open-circuit condition. The interlayer at the electrode/organic interface not only modifies the built-in potentials but affects the extraction efficiency. This can lead to a reduction in the recombination rate and hence a change in the charge collection efficiency.[47] Several reports have investigated the impact of interlayers on injection barriers, built-in fields, surface energy, and surface charge recombination rates in organic solar cells.[48–50] In consideration of these findings, it was postulated that a good solar cell interface should satisfy certain requirements.[57] These requirements include (1) promoting an ohmic contact formation between electrodes and the active layer, (2) having the appropriate energy levels to improve charge selectivity for corresponding electrodes, (3) having a large bandgap to confine excitons in the active layer, (4) possessing sufficient conductivity to reduce resistive losses, (5) having low absorption in the Vis-NIR wavelengths to minimize optical losses, (6) having chemical and physical stability to prevent undesirable reactions at the active layer/electrode interface, (7) having the ability to be processed from solution and at low temperatures, (8) being mechanically robust to support multilayer solution processing, (9) having good film-forming properties, and (10) being able to be fabricated at low cost. Indeed, this is a tall order for any material to satisfy.[51] But, there has been some hope in systematically characterizing these important requirements with the most promising materials.[51–60] For example, crosslinkable charge-transporting materials are one class of interfacial materials that have already shown their promise in controlling the charge injection and extraction properties of organic light-emitting diodes and photodetectors.[61] The HOMO/LUMO levels, bandgaps, and the hole mobility of these materials can be tuned by rational molecular design while the crosslinkers are introduced on the active components as pendant groups. As a result, these materials can be processed thermally or by photochemically crosslinking in order to form robust charge-transporting films with improved solvent resistance for the subsequent processing of the BHJ layer.

However, the first requirement that the contacts be ohmic is a major obstacle. In general, with the specific knowledge of the surface chemistry, an ohmic contact at the organic/metal interface can be achieved by interface engineering.[63,64] Typical approaches are to insert an interface layer to match the interface energy alignment or insert a dipole layer at the metal–organic interface. As stated earlier, low-work-function metals are the most common materials to form favorable energy alignment. Some reports have shown that some success can be obtained with the use of metal oxide and transition metals are also favorable materials for energy alignment at the interface, especially in inverted devices.[51] As the reader can identify in Figure 5.4, poly (3,4-ethylenedioxyhiophene) polystyrene sulfonate (PEDOT:PSS) is the most commonly used hole-transporting layer to form ohmic contacts at the anode.[62] And generally, metals such as Ca are used as discussed earlier. And as it was mentioned previously, many low-work-function metals (such as Ca) suffer from oxidation under ambient conditions; electrode degradation is a major concern for this type of device. Some researchers have tried to limit this problem by inserting another layer of material between the BHJ and metal cathode, which can avoid the use of LiF or Ca to

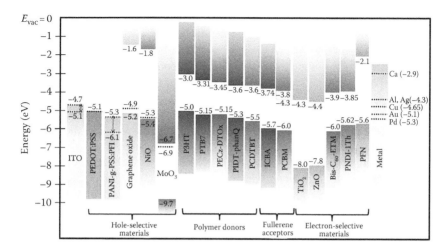

FIGURE 5.4 (**See color insert.**) Schematic view of the energy gaps and energy levels of some of the components of recent OPVs including transparent electrodes, hole-selective materials, polymer donors, fullerene acceptors, electron-selective materials and metal electrodes. The dotted lines correspond to the work functions of the materials. (From Yip, H.L. and Jen, A.K.Y., *Energy and Environmental Science*, 5(3), 5994, 2012.)

improve the interfacial stability. However, some still prefer to use metals such as Cu and Ag as cathodes when appropriate interfacial materials are applied. The convention of making devices and perfecting the art of matching layers has taken much time. Only recently do scientists and engineers know the basic steps in creating the most promising interface, which has included making self-assembled monolayers (SAMs).

The reason some researchers have utilized SAMs for creating interfacial dipoles is that the monolayer may modify the electrode work function and adjust the barrier height.[65] In order for the process of long-range charge separation to occur, there must be a driving energy for charge separation allowing the system to reach delocalized band states.[66] A high open-circuit voltage in organic photovoltaics requires an anode with a large work function to match the HOMO of the donor polymer. In many cases, the work function of PEDOT:PSS is not sufficient and modification of the electrode work function is required to facilitate efficient hole extraction. This can be realized by introducing a SAM interlayer with an intrinsic dipole at the electrode interface to control the interface energies.[67] SAMs with electron-withdrawing groups lead to protonation of the conducting substrate (such as ITO) surface, which facilitates the formation of an interfacial dipole pointing away from the electrode and increases the work function of the electrode.[68,69] It has been shown that SAMs with electron-donating groups form dipoles in the opposite direction and therefore decrease the electrode work function to be used for cathode contacts without using alkaline metal compounds.[66] For example, small molecule solar cells with two different interlayers, aluminum phthalocyanine chloride (AlPcCl) and molybdenum oxide (MoO_3) were investigated.[70] The work with these devices found that by adopting MoO_3 instead of AlPcCl for an organic solar device with the structure ITO/AlPcCl or $MoO_3/AlPcCl:C_{60}/C_{60}/LiF/Al$, the V_{OC} increases from 0.79 to 0.83 V and *FF* increases from 40% to 47.7%, which leads to an increase in PCE from 1.78% to 2.24%.[70] UPS studies found that the devices with AlPcCl as an interlayer, a vacuum level shift of 0.2 eV in the device was observed upon deposition of the first monolayer of AlPcCl and then it remained constant with further deposition.[70] In the case of the MoO_3 as an interlayer, the valence band maximum of the MoO_3 layer is 2.53 eV below the Fermi level along with a 2 eV shift in the vacuum level. The inclusion of MoO_3 leads to a shift in the HOMO level of the AlPcCl layer, where the band bending extends to about 10 nm into the AlPcCl layer and leading to the formation of a built-in field at the $MoO_3/AlPcCl$ interface.[70] It was concluded that the enhanced hole extraction due to the formation of this built-in field results in a reduction of the cell series resistance and carrier recombination, leading to an increase in both V_{OC} and *FF*. Research on interface SAMs have also been used in combination with functional metal oxides such as TiO_2 and ZnO.[71] ZnO/metal interface modification has been reported by adopting series of SAMs containing benzoic acid. It was shown that SAMs with a negative dipole moment decrease the barrier at the Al/ZnO interface and better performance.[71,72] In addition to interface energies, SAMs can also change the wettability of the substrate surface. It has been suggested that changes in wettability can affect the film morphology, which subsequently results in changes in the charge separation efficiency and the charge transport properties.[72]

EXCITON DIFFUSION AT THE INTERFACE IN ORGANIC SOLAR CELLS

In a solar cell with an organic material as the light-absorbing medium, photon absorption leads initially to the production of a strongly bound electron-hole pair, which is termed an exciton. From previous chapters, it should be clear that the excitons must diffuse to an interface with another organic material to dissociate into free charges. It is hoped that the reader appreciates the importance of this process in organic materials. The charges produced by interfacial exciton dissociation must escape from recombination, so that they can contribute to the photocurrent delivered by the solar cell. The current performance of organic-based solar cells is often to a large extent limited by the relatively short exciton diffusion length and small yield of free charge carriers (Figure 5.5).[73–77]

A great deal of research has been devoted to understanding the effects of chemical composition and material morphology on exciton diffusion and escape of charges from recombination. Time-resolved (pump-probe as described in Chapter 3) laser techniques are used to study exciton and charge dynamics in bilayered structures and heterogeneous blends of electron and hole transporting materials.[78,79] An example of such studies involves light-absorbing porphyrin molecules in combination with a wide bandgap semiconductor which acts as an electron acceptor.[80] Light absorption by the porphyrin molecule leads to the production of an exciton. The exciton may diffuse toward the interface with the semiconductor and decay by the injection of an electron into the conduction band.[81] The distance an exciton is able to cover by diffusion is, for most organic dye layers, typically of the order of only 1–5 nm. This

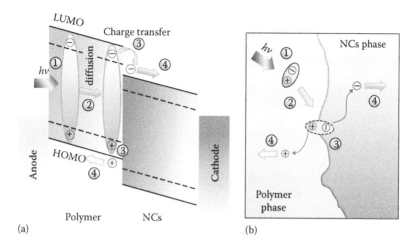

(a)   (b)

FIGURE 5.5 **(See color insert.)** Schematic diagram of the photocurrent generation mechanism in bulk heterojunction hybrid solar cells. (a) Schematics of the different energy levels and the individual processes: exciton generation (1), exciton diffusion (2), charge transfer (3), charge carrier transport, and collection (4). (b) Schematic of a bulk heterojunction contact at the polymer–NC interface. The same processes mentioned in (a) are illustrated directly at the donor–acceptor interface. (From Zhou, Y.F. et al., *Energy Environ. Sci.*, 3(12), 1851, 2010.)

diffusion length corresponds to a few hopping steps of the exciton prior to its decay back to the ground state. Indeed, over the last two decades, researchers have found that the exciton diffusion length via interfacial considerations can be significantly enhanced by modification of the molecular system. For example, a longer diffusion length could be realized by the introduction of a Pd atom in the central core of porphyrin molecules.[82] The heavy Pd atoms lead to the conversion of singlet excitons to triplet excitons. The long lifetime of the triplets leads to a diffusion length, which is ten times longer than in the case when hydrogen atoms are present in the core of the porphyrin molecule (Figure 5.6).[82]

There have been investigations of nondispersive exciton diffusion for particular solar cells. The impact of this work stems from the fact that in unordered polymers, the electronically excited states of the segmental sites exhibit a large energy distribution. Excitons migrate to energetically favored sites until they are trapped at the site with the lowest excitation energy of the local environment.[83] In order to limit this effect, scientists have investigated exciton diffusion in polymer matrices highly doped with dye molecules, which are the active sites for exciton migration. If they interact weakly with the environment, they should have a narrow energy distribution and trapping should be reduced. In this case, exciton diffusion should be essentially loss-free and large diffusion lengths are possible. However, very large diffusion lengths have not been observed at this point in typical conjugated polymers. There has been some success in other macromolecular architectures.[83]

For organic devices, film thicknesses on the order of 100–200 nm are typically required to absorb most of the incident light due to the high absorption coefficient

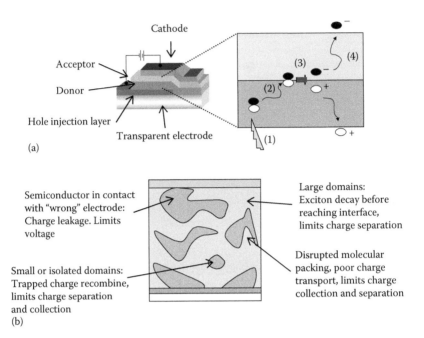

FIGURE 5.6 **(See color insert.)** (a) Schematic of a bilayer organic solar cell illustrating the main processes occurring in the photoactive layer: (1) absorption of photon to create an exciton, (2) exciton diffusion, (3) exciton splitting at the interface between donor and acceptor, and (4) diffusion and collection of charges. (b) Schematic of the side-view microstructure of a bulk heterojunction donor–acceptor blend film, illustrating the ways in which a non-optimum microstructure can affect device performance. (From Brabec, C.J. et al., *Chem. Soc. Rev.*, 40(3), 1185, 2011; Shaw, P.E. et al., *Adv. Mater.*, 20, 3516, 2008.)

of conjugated materials. Thus, for a simple bilayer device, this would result in the majority of the excitons being created away from the donor/acceptor interface and they ultimately would not be harvested. It should be noted that for devices based upon highly crystalline, vacuum-sublimed organic semiconductors, the exciton diffusion lengths are longer, resulting in good efficiencies even for bilayer devices. One solution to this problem for soluble materials is to mix the donor and acceptor in a BHJ. With appropriate control of the morphology, all excitons can be created within a diffusion length of a donor/acceptor interface and hence be harvested.[83] Therefore, control of the nanoscale morphology, or microstructure of the blend is critical to ensuring that all excitons are collected and dissociated. Once the exciton has dissociated, the free holes and electrons must then be transported through the donor and acceptor phases to their respective electrodes. Consequently, continuous percolation pathways are required through each phase. In addition, the transport of the hole and electron through each phase will be strongly influenced by the local order and crystallinity within each phase, with crystalline morphologies generally leading to the highest charge carrier mobilities.

SURFACE CONTROL ON INTERFACIAL
DIPOLES IN ORGANIC SOLAR CELLS

The strategies mentioned earlier mainly concern the control of the microstructure of the donor–acceptor blend film in order to achieve the optimum combination of domain size and connectivity. In this way, one can maximize charge pair generation and current collection. However, efficient photocurrent generation in a bulk heterojunction also depends upon the existence of selective electrodes, each of which permits the barrier-free collection of one charge type and blocks the other, in order to direct the photocurrent. The photocurrent can also be assisted by the vertical composition profile of the photoactive layer.[83] For example, a buffer layer of pure donor beside the anode would minimize electron leakage to the anode, while a layer of acceptor close to the cathode would block hole leakage. A vertical composition gradient would assist in driving the hole current toward the donor-rich side and the electron current toward the acceptor-rich side, through the associated gradient in conductivity and density of states.

For evaporated molecular active layers, a vertical gradient in composition can be built in through the successive co-deposition of blend layers of different composition. This approach has been shown to improve fill factor and open-circuit voltage in copper phthalocyanine:C60 heterojunctions. For solution-processed active layers, the successive processing of polymer layers is difficult and requires orthogonal solvents or solid film transfer techniques. An alternative approach is to rely upon the different interaction energies between the components and the substrate or the top surface to drive the preferential segregation of the different components toward or away from either interface. Researchers have shown that vertical phase segregation could be achieved in a blend of poly(9,9′-dioctylfluorene-co-bis-N,N′-(4-butylphenyl)-bis-N,N′-phenyl-1,4-phenylenediamine) (PFB) and poly(9,9′-dioctylfluoreneco-benzothiadiazole) (F8BT) when the components were cast from a high boiling point solvent and that more complete vertical segregation could be achieved by coating the substrate with a silane-based molecular monolayer to reduce its surface energy.[84] There have also been scientists that have shown that the choice of a high boiling point solvent increases the extent of stratification within a polyfluorene:PCBM blend film and influences the device performance.[83]

While there are many things one can do to reduce the issues observed at the interface, the problem of phase segregation in organic polymers still seems to be an issue that limits the success of most approaches. The phase segregation process has a large impact on the surface energy of the polymeric solar cell. There has been evidence for spontaneous vertical phase segregation in blends of polymers with PCBM.[85] It was shown by using spectroscopic ellipsometry that P3HT tends to segregate toward the top of a 1:1 blend film of P3HT and PCBM and that this segregation is enhanced following thermal annealing for a blend film on silica.[86] However, replacing the silica substrate with a lower surface energy substrate led to the opposite sense of segregation, with P3HT moving toward the substrate. These studies confirmed the different sense of segregation on high and low energy substrates (OTS and SiO_2, respectively) by probing the top and buried interfaces of a P3HT:PCBM blend film using NEXAFS on delaminated films. When applied onto a low surface energy substrate, the blend component with the lower surface energy was shown to migrate to the interface in order to minimize the interfacial free energy, which is proportional to the difference between the surface energies of the blend and substrate. When a high energy substrate surface such as silicon dioxide is employed, the higher surface energy PCBM component preferentially segregates to this surface.[86] In all cases, the P3HT migrates preferentially to the air interface to create the lowest energy surface. As the reader can see, this process is rather complex and to a great extent some of the approaches have taken a structure-function correlation method, which requires the synthesis of many potential candidates. Here, as well as the rather long list of requirements mentioned earlier, is where new efforts in interfacial engineering are gaining new interest with intensified efforts.

The interface is a major obstacle in the production of all organic solar cells with higher efficiencies. The large number of requirements for organic interfaces makes the milestone of reaching the efficiency found for silicon appear insurmountable. Perhaps before this milestone is obtained, new scientific processes may need to be detailed with further investigations. It is not surprising that through all the excellent investigations covered thus far in this book, the problem of the interface turns out to be the most difficult. Some have suggested that this stems from the construction (or geometry) of the organic solar cell itself. While organic polymers are great in their ability to be tailored synthetically and produced at high volume, they do suffer from problems of long range order or regions of microstructural order only. In an organic solar cell, there can be clusters of donor and acceptor regions separated and regions where the two are mixed. As we have discussed earlier, these clusters of donor and acceptors (often spin coated on substrates) and the subsequent mixed regions can vary in size. Here lies the problem. Presently, the effect of the mixed domains has not been properly understood in the case of organic polymeric solar cells. The effect of these domains on charge collections at the electrodes is still a matter of great research interest.

REFERENCES

1. Y. Yin and A.P. Alivisatos, Colloidal nanocrystal synthesis and the organic-inorganic interface, *Nature*, 437(7059), 664–670, 2005.
2. A.P. Alivisatos, Semiconductor clusters, nanocrystals, and quantum dots, *Science*, 271(5251), 933–937, 1996.
3. W.L. Ma, C.Y. Yang, X. Gong, K. Lee, and A.J. Heeger, Thermally stable, efficient polymer solar cells with nanoscale control of the interpenetrating network morphology, *Advanced Functional Materials*, 15(10), 1617, 2005.
4. H.P. Zhou, Q. Chen, G. Li, S. Luo, T.B. Song, H.S. Duan, Z.R. Hong, J.B. You, Y.S. Liu, and Y. Yang, Interface engineering of highly efficient perovskite solar cells, *Science*, 345(6196), 542, 2014.
5. P.V. Kamat, Photophysical, photochemical and photocatalytic aspects of metal nanoparticles, *Journal of Physical Chemistry B*, 106(32), 7729, 2002.
6. D. Beljonne, J. Cornil, L. Muccioli, C. Zannoni, J.L. Bredas, and F. Castet, Electronic processes at organic-organic interfaces: Insight from modeling and implications for opto-electronic devices, *Chemistry of Materials*, 23(3), 591, 2011.
7. M. Scharber, D. Mühlbacher, M. Koppe, P. Denk, C. Waldauf, A. Heeger, and C. Brabec, Design rules for donors in bulk-heterojunction solar cells—Towards 10% energy-conversion efficiency, *Advanced Materials*, 18(6), 789–794, 2006.

8. K.R. Graham et al., Importance of the donor:fullerene intermolecular arrangement for high-efficiency organic photovoltaics, *Journal of the American Chemical Society*, 136(27), 9608, 2014.

9. S. Braun, W.R. Salaneck, and M. Fahlman, Energy-level alignment at organic/metal and organic/organic interfaces, *Advanced Materials*, 21(14), 1450, 2009.

10. M.T. Greiner, M.G. Helander, W.M. Tang, Z.B. Wang, J. Qiu, and Z.H. Lu, Universal energy-level alignment of molecules on metal oxide, *Nature Materials*, 11(1), 76, 2012.

11. S. Zhong, J.Q. Zhong, H.Y. Mao, J.L. Zhang, J.D. Lin, and W. Chen, The role of gap states in the energy level alignment at the organic–organic heterojunction interfaces, *Physical Chemistry Chemical Physics*, 14, 14127–14141, 2012.

12. M. Oehzelt, N. Koch, and G. Heimel, Organic semiconductor density of states controls the energy level alignment at electrode interfaces, *Nature Communications*, 5, 4174, 2014.

13. P.K. Nayak, K.L. Narasimhan, and D. Cahen, Separating charges at organic interfaces: Effects of disorder, hot states, and electric field, *The Journal of Physical Chemistry Letters*, 4, 1707, 2013.

14. C. Schwarz, S. Tscheuschner, J. Frisch, S. Winkler, N. Koch, H. Bassler, and A. Kohler, Role of the effective mass and interfacial dipoles on exciton dissociation in organic donor-acceptor solar cells, *Physical Review B*, 87(15), 155205, 2013.

15. S.-S. Sun and N.S. Sariciftci, *Organic Photovoltaics: Mechanisms, Materials, and Devices*, p. 427, CRC Press, Boca Ration, FL, March 29, 2005.

16. H. Xu, R. Chen, Q. Sun, W. Lai, Q. Su, W. Huang, and X. Liu, Recent progress in metal–organic complexes for optoelectronic applications, *Chemical Society Reviews*, 43, 3259, 2014.

17. M.R. Krcmar and W.M. Saslow, Model for electrostatic screening by a semiconductor with free surface carriers, *Physical Review B*, 66(23), 235310, 2002.

18. J.-L. Bredas, D. Beljonne, V. Coropceanu, and J. Cornil, Charge-transfer and energy-transfer processes in π-conjugated oligomers and polymers: A molecular picture, *Chemical Reviews*, 104, 4971, 2004.

19. Y. Huang, S. Westenhoff, I. Avilov, P. Sreearunothai, J.M. Hodgkiss, C. Deleener, R.H. Friend, and D. Beljonne, Electronic structures of interfacial states formed at polymeric semiconductor heterojunctions, *Nature Materials*, 7, 483, 2008.

20. Y. Zhou, J. Pei, Q. Dong, X. Sun, Y. Liu, and W. Tian, Donor–acceptor molecule as the acceptor for polymer-based bulk heterojunction solar cells, *The Journal of Physical Chemistry C*, 113, 7882, 2009.

21. O.P. Varnavski, L. Sukhomlinova, R. Tweig, G.C. Bazan, T. Goodson III, Coherent effects in energy transport in model dendritic structures investigated by ultra-fast fluorescence anisotropy spectroscopy, *Journal of the American Chemical Society*, 124, 1736–1743, 2002.

22. R. West, S. Lahankhar, H.B. Xie, O. Varnavski, M. Ranasinghe, and T. Goodson III, Electronic coupling in organic branched macromolecules investigated by nonlinear optical and fluorescence upconversion spectroscopy, *The Journal of Chemical Physics*, 120(1), 337–344, 2004.

23. M. Ransinghe, M.W. Hager, C.B. Gorman, and T. Goodson III, Energy transfer in phenylacetylene dendrimers revisited, *Journal of Physical Chemistry*, 108, 8543–8549, 2004.

24. T. Goodson III, Optical excitations in novel organic branched structures investigated by time-resolved and nonlinear optical spectroscopy, *Accounts of Chemical Research*, 38, 99–107, 2005.

25. V. Coropceanu, J. Cornil, D.A. da Silva, Y. Olivier, R. Silbey, and J.L. Bredas, Charge transport in organic semiconductors, *Chemical Reviews*, 107(4), 926, 2007.

26. H. Bassler and A. Keohler, Charge transport in organic semiconductors, *Topics in Current Chemistry*, 312, 1–66, 2012.

27. H. Ohkita et al., Charge carrier formation in polythiophene/fullerene blend films studied by transient absorption spectroscopy, *Journal of the American Chemical Society*, 130(10), 3030, 2008.

28. T.M. Clarke and J.R. Durrant, Charge photogeneration in organic solar cells, *Chemical Reviews*, 110(11), 6736, 2010.

29. C. Deibel, T. Strobel, and V. Dyakonov, Role of the charge transfer state in organic donor-acceptor solar cells, *Advanced Materials*, 22(37), 4097, 2010.

30. H. Ohkita and S. Ito, Transient absorption spectroscopy of polymer-based thin-film solar cells, *Polymer*, 52(20), 4397, 2011.

31. A.A. Bakulin, A. Rao, V.G. Pavelyev, P.H.M. van Loosdrecht, M.S. Pshenichnikov, D. Niedzialek, J. Cornil, D. Beljonne, and R.H. Friend, The role of driving energy and delocalized states for charge separation in organic semiconductors, *Science*, 335(6074), 1340, 2012.

32. K.B. Ornso, E.O. Jonsson, K.W. Jacobsen, and K.S. Thygesen, Importance of the reorganization energy barrier in computational design of porphyrin-based solar cells with cobalt-based redox mediators, *The Journal of Physical Chemistry C*, 119(23), 12792, 2015.

33. H. Imahori and S. Fukuzumi, Porphyrin- and fullerene-based molecular photovoltaic devices, *Advanced Functional Materials*, 14, 525, 2004.

34. B. Yang, Y. Yi, C.-R. Zhang, S.G. Aziz, V. Coropceanu, and J.-L. Brédas, Impact of electron delocalization on the nature of the charge-transfer states in model pentacene/C60 interfaces: A density functional theory study, *Journal of Physical Chemistry C*, 118(48), 27648–27656, 2014.

35. J. Shi, X. Xu, D. Li, and Q. Meng, Interfaces in perovskite solar cells, *Small*, 11(21), 2472–2486, 2015.

36. S. Braun, X. Liu, W.R. Salaneck, and M. Fahlman, Fermi level equilibrium at donor-acceptor interfaces in multi-layered thin film stack of TTF and TCNQ, *Organic Electronics*, 11(2), 212, 2010.

37. Q.Y. Bao, O. Sandberg, D. Dagnelund, S. Sanden, S. Braun, H. Aarnio, X.J. Liu, W.M.M. Chen, R. Osterbacka, and M. Fahlman, Trap-assisted recombination via integer charge transfer states in organic bulk heterojunction photovoltaics, *Advanced Functional Materials*, 24(40), 6309, 2014.

38. M. Fahlman, A. Crispin, X. Crispin, S.K.M. Henze, M.P. de Jong, W. Osikowicz, C. Tengstedt, and W.R. Salaneck, Electronic structure of hybrid interfaces for polymer-based electronics, *Journal of Physics: Condensed Matter*, 19(18), 183202, 2007.

39. H. Vazquez, R. Oszwaldowski, P. Pou, J. Ortega, R. Perez, F. Flores, and A. Kahn, Dipole formation at metal/PTCDA interfaces: Role of the charge neutrality level, *Europhysics Letters*, 65(6), 802, 2004.

40. M.G. Mason et al., Interfacial chemistry of Alq(3) and LiF with reactive metals, *Journal of Applied Physics*, 89(5), 2756, 2001.

41. (a) P.S. Davids and D.L. Smith, Non-degenerate continuum model for polymer-light emitting diodes, *Journal of Applied Physics*, 78, 4244–4252, 1995; (b) T.M. Brown, R.H. Friend, I.S. Millard, D.J. Lacey, T. Butler, J.H. Burroughs, and F. Cacialli, Electronic line-up in light emitting diodes with alkali-halide metal cathodes, *Journal of Applied Physics*, 77, 3096, 2000.

42. Z. Xu, L.-M. Chen, M.-H. Chen, G. Li, and Y. Yang, Energy level alignment of poly(3-hexylthiophene): [6,6]-phenyl C61C61 butyric acid methyl ester bulk heterojunction, *Applied Physics Letters*, 95, 013301, 2009.

43. T. Kugler, M. Logdlund, and W.R. Salaneck, Photoelectron spectroscopy and quantum chemical modeling applied to polymer surfaces and interfaces in light-emitting devices, *Accounts of Chemical Research*, 32(3), 225, 1999.

44. Z. Xu, L.M. Chen, G.W. Yang, C.H. Huang, J.H. Hou, Y. Wu, G. Li, C.S. Hsu, and Y. Yang, Vertical phase separation in poly(3-hexylthiophene): Fullerene derivative blends and its advantage for inverted structure solar cells, *Advanced Functional Materials*, 19(8), 1227, 2009.
45. A.L. Ayzner, C.J. Tassone, S.H. Tolbert, and B.J. Schwartz, Reappraising the need for bulk heterojunctions in polymer-fullerene photovoltaics: The role of carrier transport in all-solution-processed P3HT/PCBM bilayer solar cells, *Journal of Physical Chemistry C*, 113(46), 20050, 2009.
46. F. Etzold, I.A. Howard, N. Forler, D.M. Cho, M. Meister, H. Mangold, J. Shu, M.R. Hansen, K. Mullen, and F. Laquai, The effect of solvent additives on morphology and excited-state dynamics in PCPDTBT:PCBM photovoltaic blends, *Journal of the American Chemical Society*, 134(25), 10569, 2012.
47. M. Campoy-Quiles, T. Ferenczi, T. Agostinelli, P.G. Etchegoin, Y. Kim, T.D. Anthopoulos, P.N. Stavrinou, D.D.C. Bradley, and J. Nelson, Morphology evolution via self-organization and lateral and vertical diffusion in polymer:fullerene solar cell blends, *Nature Materials*, 7, 158–164, 2008.
48. A. Zhugayevych and S. Tretiak, Theoretical description of structural and electronic properties of organic photovoltaic materials, *Annual Review of Physical Chemistry*, 66, 305, 2015.
49. Y. Li, Z.Y. Fu, and B.L. Su, Hierarchically structured porous materials for energy conversion and storage, *Advanced Functional Materials*, 22(22), 4634, 2012.
50. A. Wagenpfahl, C. Deibel, and V. Dyakonov, Organic solar cell efficiencies under the aspect of reduced surface recombination velocities, *IEEE Journal of Selected Topics in Quantum Electronics*, 16(6), 1759, 2010.
51. H.L. Yip and A.K.Y. Jen, Recent advances in solution-processed interfacial materials for efficient and stable polymer solar cells, *Energy and Environmental Science*, 5(3), 5994, 2012.
52. K. Walzer, B. Maennig, M. Pfeiffer, and K. Leo, Highly efficient organic devices based on electrically doped transport layers, *Chemistry Review*, 107(4), 1233–1271, 2007.
53. A.W. Hains, C. Ramanan, M.D. Irwin, J. Liu, M.R. Wasielewski, and T.J. Marks, Designed bithiophene-based interfacial layer for high-efficiency bulk-heterojunction organic photovoltaic cells, Importance of interfacial energy level matching, *ACS Applied Materials and Interfaces*, 2, 175–185, 2010.
54. V. Shrotriya, G. Li, Y. Yao, C.W. Chu, and Y. Yang, Transition metal oxides as the buffer layer for polymer photovoltaic cells, *Applied Physics Letters*, 88, 073508, 2006.
55. Z.W. Gu, L.J. Zuo, T.T. Larsen-Olsen, T. Ye, G. Wu, F.C. Krebs, and H.Z. Chen, Interfacial engineering of self-assembled monolayer modified semi-roll-to-roll planar heterojunction perovskite solar cells on flexible substrates, *Journal of Materials Chemistry A*, 3(48), 24254, 2015.
56. Z.H. Wu, C. Sun, S. Dong, X.F. Jiang, S.P. Wu, H.B. Wu, H.L. Yip, F. Huang, and Y. Cao, n-Type water/alcohol-soluble naphthalene diimide-based conjugated polymers for high-performance polymer solar cells, *Journal of the American Chemical Society*, 138(6), 2016, 2004.
57. C. Sun et al., Amino-functionalized conjugated polymer as an efficient electron transport layer for high-performance planar-heterojunction perovskite solar cells, *Advanced Energy Materials*, 6(5), 1501534, 2016.
58. W.J. Potscavage, A. Sharma, and B. Kippelen, Critical interfaces in organic solar cells and their influence on the open-circuit voltage, *Accounts of Chemical Research*, 42(11), 1758, 2009.
59. G. Heimel, L. Romaner, E. Zojer, and J.L. Bredas, The interface energetics of self-assembled monolayers on metals, *Accounts of Chemical Research*, 41(6), 721, 2008.
60. S. Murase and Y. Yang, Solution processed MoO_3 interfacial layer for organic photovoltaics prepared by a facile synthesis method, *Advanced Materials*, 24(18), 2459, 2012.

61. J.M. Yun, J.S. Yeo, J. Kim, H.G. Jeong, D.Y. Kim, Y.J. Noh, S.S. Kim, B.C. Ku, and S.I. Na, Solution-processable reduced graphene oxide as a novel alternative to PEDOT:PSS hole transport layers for highly efficient and stable polymer solar cells, *Advanced Materials*, 23(42), 4923, 2011.

62. P.W.M. Blom, V.D. Mihailetchi, L.J.A. Koster, and D.E. Markov, Device physics of polymer: Fullerene bulk heterojunction solar cells, *Advanced Materials*, 19(12), 1551, 2007.

63. H. Ishii, K. Sugiyama, E. Ito, and K. Seki, Energy level alignment and interfacial electronic structures at organic/metal and organic/organic interfaces, *Advanced Materials*, 11(8), 605, 1999.

64. M. Kroger, S. Hamwi, J. Meyer, T. Riedl, W. Kowalsky, and A. Kahn, Role of the deep-lying electronic states of MoO_3 in the enhancement of hole-injection in organic thin films, *Applied Physics Letters*, 95(12), 123301, 2009.

65. C. Tengstedt, W. Osikowicz, W.R. Salaneck, I.D. Parker, C.H. Hsu, and M. Fahlman, Fermi-level pinning at conjugated polymer interfaces, *Applied Physics Letters*, 88(5), 053502, 2006.

66. T.H. Lai, S.W. Tsang, J.R. Manders, S. Chen, and F. So, Properties of interlayer for organic photovoltaics, *Materials Today*, 16(11), 424, 2013.

67. F. Rissner, G.M. Rangger, O.T. Hofmann, A.M. Track, G. Heimel, and E. Zojer, Understanding the electronic structure of metal/SAM/organic: Semiconductor hetero-junctions, *ACS Nano*, 3(11), 3513, 2009.

68. A.M. Nardes, M. Kemerink, M.M. de Kok, E. Vinken, K. Maturova, and R.A.J. Janssen, Conductivity, work function, and environmental stability of PEDOT:PSS thin films treated with sorbitol, *Organic Electronics*, 9(5), 727, 2008.

69. S. van Reenen, S. Kouijzer, R.J. Janssen, M.M. Wienk, and K. Martijn, origin of work function modification by ionic and amine-based interface layers, *Advanced Materials Interfaces*, 1(8), 1400189, 2014.

70. T.A. Papadopoulos, J. Meyer, H. Li, Z.L. Guan, A. Kahn, and J.L. Bredas, Nature of the interfaces between stoichiometric and under-stoichiometric MoO_3 and 4,4″-N,N′-dicarbazole-biphenyl: A combined theoretical and experimental study, *Advanced Functional Materials*, 23(48), 6091, 2013.

71. (a) H.L. Yip, S.K. Hau, N.S. Baek, H. Ma, and A.K.Y. Jen, Polymer solar cells that use self-assembled monolayer modified ZnO/metals as cathodes, *Advanced Material*, 20(12), 2376–2382, 2008; (b) G. Li, C.W. Chu, V. Shrotriya, J. Huang, and Y. Yang, Efficient inverted polymer solar cells, *Applied Physics Letters*, 88(25), 253503, 2006.

72. D.S. Germack, C.K. Chan, B.H. Hamadani, L.J. Richter, D.A. Fischer, D.J. Gundlach, and D.M. DeLongchamp, Substrate-dependent interface composition and charge transport in films for organic photovoltaics, *Applied Physics Letters*, 94(23), 233303, 2009.

73. Y. Terao, H. Sasabe, and C. Adachi, Correlation of hole mobility, exciton diffusion length, and solar cell characteristics in phthalocyanine/fullerene organic solar cells, *Applied Physics Letters*, 90(10), 103515, 2007.

74. G.C. Xing, N. Mathews, S.Y. Sun, S.S. Lim, Y.M. Lam, M. Gratzel, S. Mhaisalkar, and T.C. Sum, Long-range balanced electron- and hole-transport lengths in organic-inorganic $CH_3NH_3PbI_3$, *Science*, 342(6156), 344, 2013.

75. H. Najafov, B. Lee, Q. Zhou, L.C. Feldman, and V. Podzorov, Observation of long-range exciton diffusion in highly ordered organic semiconductors, *Nature Materials*, 9(11), 938, 2010.

76. S.R. Scully and M.D. McGehee, Effects of optical interference and energy transfer on exciton diffusion length measurements in organic semiconductors, *Journal of Applied Physics*, 100(3), 034907, 2006.

77. Y.F. Zhou, M. Eck, and M. Kruger, Bulk-heterojunction hybrid solar cells based on colloidal nanocrystals and conjugated polymers, *Energy and Environmental Science*, 3(12), 1851, 2010.

78. A.E. Jailaubekov et al., Hot charge-transfer excitons set the time limit for charge separation at donor/acceptor interfaces in organic photovoltaics, *Nature Materials*, 12(1), 66, 2013.

79. X.Y. Zhu, Q. Yang, and M. Muntwiler, Charge-transfer excitons at organic semiconductor surfaces and interfaces, *Accounts of Chemical Research*, 42(11), 1779, 2009.

80. M.J. Griffith, K. Sunahara, P. Wagner, K. Wagner, G.G. Wallace, D.L. Officer, A. Furube, R. Katoh, S. Mori, and A.J. Mozer, Porphyrins for dye-sensitised solar cells: New insights into efficiency-determining electron transfer steps, *Chemical Communication*, 48(35), 4145, 2012.

81. L.L. Li and E.W.G. Diau, Porphyrin-sensitized solar cells, *Chemical Society of Review*, 42(1), 291, 2013.

82. A. Huijser, T.J. Savenije, A. Kotlewski, S.J. Picken, and L.D.A. Siebbeles, Efficient light-harvesting layers of homeotropically aligned porphyrin derivatives, *Advanced Materials*, 18(17), 2234, 2006.

83. (a) C.J. Brabec, M. Heeney, I. McCulloch, and J. Nelson, Influence of blend microstructure on bulk heterojunction organic photovoltaic performance, *Chemical Society of Review*, 40(3), 1185, 2011. (b) P.E. Shaw, A. Ruseckas, and I.D.W. Samuel, Exciton diffusion measurements in poly(3-hexylthiophene), *Advanced Material*, 20, 3516–3520, 2008. (c) C.M. Björström, J. Rysz, A. Benasik, A. Budkowski, F. Zhang, O. Inganäs, M.R. Andersson, K.O. Magnusson, J.J. Benson-Smith, J. Nelson, and E. Moons, Device performance of APFO-3/PCBM solar cells with controlled morphology, *Advanced Materials*, 21(43), 4398–4403, 2009.

84. M. Abdulla, C. Renero-Lecuna, J.S. Kim, and R.H. Friend, Morphological study of F8BT:PFB thin film blends, *Organic Electronics*, 23, 87, 2015.

85. J.K.J. van Duren, X.N. Yang, J. Loos, C.W.T. Bulle-Lieuwma, A.B. Sieval, J.C. Hummelen, and R.A.J. Janssen, Relating the morphology of poly(p-phenylene vinylene)/methanofullerene blends to solar-cell performance, *Advanced Functional Materials*, 14(5), 425, 2004.

86. G.Y. Zhang, S.A. Hawks, C. Ngo, L.T. Schelhas, D.T. Scholes, H. Kang, J.C. Aguirre, S.H. Tolbert, and B.J. Schwartz, Extensive penetration of evaporated electrode metals into fullerene films: Intercalated metal nanostructures and influence on device architecture, *ACS Applied Materials and Interfaces*, 7(45), 25247, 2015.

6 New Approaches: Testing the Limits in Organic Solar Cells

INTRODUCTION

The use of organic materials for the construction of bulk heterojunction solar cells has indeed developed substantially since the first reports appeared in the literature. While the efficiency of these devices continues to rise, perhaps even a more impressive development is the level of detailed understanding of the physics of the organic materials used in the devices. There is still a great hope for further development of these materials and devices. As it was illustrated in the previous chapter, there are new materials and new device constructions that are rigorously being investigated. The use of new synthetic strategies toward molecules (both small and large) may aid in the development of superior organic materials with enhanced absorption and longer exciton diffusion lengths.[2] Additionally, new ideas regarding the formation of molecular aggregates have suggested even further enhancement in the exciton diffusion process, which is of vital importance to the efficiency of an organic solar cell. There have been recent reports of the formation of ordered organic arrays (by π–π stacking) that may give rise to exciton diffusion lengths on the order of ~1 μm.[3] It is hoped that this would be a demonstrative enhancement over previous organic solar material results. The potential of these new approaches certainly would warrant further investigations (and investments) into this area of research with organic solar cells.

The future does look bright for organic materials. And perhaps what might give the focus of organic solar cells more leverage is that recently there are alternative approaches to using the conventional bulk heterojunction organic solar cell devices. These approaches still contain organic parts in their core technology; however, the mechanism of their operation is substantially different. In this chapter, we discuss two of these new approaches: (1) multiple exciton generation and singlet exciton fission (Figure 6.1) and (2) the use of organometallic perovskites. The formation of materials for the purpose of enhancing the overall efficiency of a device will be discussed as well as the current understanding of important mechanisms of the two processes. This chapter serves as an introduction to these two exciting and new areas related to organic and organometallic solar cells. Indeed, a whole volume could be devoted to the details of these subjects alone. However, it is interesting to note the similarities in the discovery and analysis process of these new materials with previously mentioned bulk heterojunction small molecule and conjugated polymer solar cells.

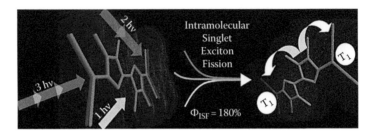

FIGURE 6.1　**(See color insert.)** Depiction of singlet exciton fission in an organic molecule.

SINGLET EXCITON FISSION

As mentioned a number of times in previous chapters, the number of carriers at the interface (and the subsequent transfer) will ultimately determine the efficiency of the solar device. There have been attempts to create more carriers in solar devices. Researchers have reported new approaches of carrier multiplication for increasing the photocurrent of solar cells.[4–7] The process of carrier multiplication was found to be very efficient in quantum-confined semiconductor nanocrystals.[8] Researchers have shown that the absorption of a single photon may produce two or even three electron–hole pairs. This could result in internal quantum efficiencies greater than 200% for the case of PbSe quantum-confined nanocrystals.[9] Indeed, this process took many researchers by surprise. The opportunity to generate multiple excitons with just one photon of light suggested that a new approach might provide further enhancement of solar cells.[10] Intensified research ensued over this impressive process. As it can be observed in Figure 6.2 from the energy diagrams based on allowed exciton states in various inorganic semiconductor nanocrystals, various electron–hole populations may be possible from the initial excitation of the higher (in energy) absorption states leading to multiple excitons being generated. It has also been reported that for CdSe, efficiencies upwards of ~165% have been observed, which is significant as these quantum-confined structures are characterized by energy diagrams and carrier dynamics that are distinctly different from those of the lead salt nanocrystals.[11]

However, there were discrepancies in some of the reports of semiconductors demonstrating the multi-exciton generation (MEG) effect, and more importantly, there were some concerns regarding the actual measurements used in these reports. Some believed that the pulsed laser experiments used to measure the MEG effect in certain inorganic semiconductors may produce a steady-state population of charged quantum dots by photoionization if the lifetime of the photoionized state is longer than the time between excitation pulses.[5] The proposed scheme is illustrated in Figure 6.2. While the details of this important process are far beyond the scope of this chapter, one can appreciate from this figure the process of an unexcited quantum dots being excited to a hot exciton state produced by absorption at twice the ground state energy gap which can either cool to the bandgap energy, thus creating a single exciton in its lowest-energy state, or it may undergo MEG to create two or more excitons.

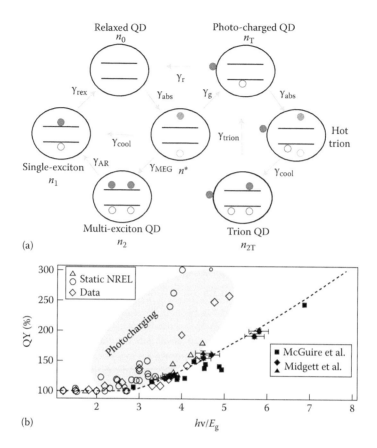

FIGURE 6.2 (See color insert.) The mechanism and quantum yield dependence in semiconductor multi-exciton generation processes. (a) Relaxation pathways for excitons produced with high-energy photons. The photocharging pathway complicates the analysis of MEG yields and depends on the photon energy, photon fluence, and QD surface quality. The effects of photocharging on MEG yields can be reduced by flowing or stirring the samples during the experiment. (b) Photon-to-exciton QYs deduced from transient absorption data and results obtained on static solutions compared to flowing or stirring solutions.

But the suddle detail in the analysis (also shown in Figure 6.2) is that it is believed that the hot exciton state may also undergo photocharging. Additional absorption of the quantum dot by the next laser pulse could result in a state, which undergoes a nonradiative recombination process, which resembles the biexciton recombination. Controversy around this particular recombination process has suggested that these experiments may lead to an artificially high QY. New experiments are now underway to strengthen the claims of the very high yields in quantum dots, which exhibit MEG processes. As mentioned in previous chapters, the use of ultrafast interfacial electron processes to separate electrons and holes generated in quantum dots is one important approach in understanding these processes.

SINGLET EXCITON FISSION IN ORGANIC MATERIALS

While many of the first reports utilized inorganic quantum dot structures, it wasn't long before those interested in organic solar cell devices started to investigate the possibility of multiple exciton generation in organic semiconducting materials.[12–20] In this case, the process most commonly invoked is that of single exciton fission. Singlet fission is a spin-allowed process in which two triplets match the energy of one singlet and subsequently lead to the formation of a triplet pair. It is often said that the two resulting triplet excitations produced from an excited singlet are born coupled into a pure singlet state. Singlet fission can therefore be viewed as a special case of internal conversion (radiationless transition between two electronic states of equal multiplicity).[20] This process is similar to many other internal conversion processes. It can be very fast, particularly in molecular crystals. When this process is isoergic or slightly exoergic and the coupling is favorable, the transformation occurs on a ps or even sub-ps time scale.[15]

While the field of organic solar cells seemed to catch on to this process with new and productive materials, it appears that the possibility of singlet fission in organic structures had already been around for some time. In fact, it was first used to probe the optical properties of well-known molecular crystals such as anthracene and tetracene, many years ago (Figure 6.3).[18] This led to the first rational diagrams of the singlet fission process in organic materials such as that in the diagram shown in Figure 6.3.[18,19] One observes the competing process of internal conversion and triplet–triplet annihilation that are critical to the success and failure of observing this process in organic materials.[18,19] But the discovery of organic materials that demonstrate this process did take some time. The use of simple conjugated molecules was not an easy solution to finding suitable materials. Additionally, it was found that the singlet exciton fission process could proceed by both an intermolecular and intramolecular approach with a variety of organic materials.[20–22]

In general, the most thoroughly investigated singlet exciton fission organic material is pentacene. This organic molecule is an example of an acene, which has five benzene rings. A number of very important reports have probed the efficiency of the singlet exciton fission process in pentacene and other acene crystals and have

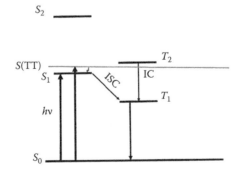

FIGURE 6.3   The energy level diagram for singlet exciton fission.

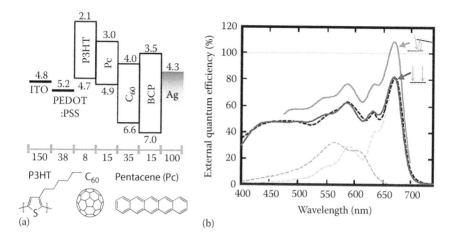

FIGURE 6.4 (See color insert.) Organic solar cell schematic as well as external quantum efficiency with pentacene as the active layer. (a) Chemical structures and architecture of the solar cell with the thickness of each layer in nanometers and energy levels of the lowest unoccupied and highest occupied molecular orbitals in electron volts [12,18,20,29–31]. The anode is composed of indium tin oxide (ITO) and poly(3,4-ethylenedioxythiophene) poly(styrenesulfonate) (PEDOT:PSS). The cathode employs bathocuproine (BCP) and a silver cap. (b) External quantum efficiency of devices without optical trapping (blue line), and device measured with light incident at 10° from normal with an external mirror reflecting the residual pump light (red line). Optical fits from IQE modeling are shown with dashed lines: modeled pentacene EQE (blue dashes), modeled P3HT EQE (purple dashes), and modeled device EQE (black dashes) for comparison to the measured device efficiency without optical trapping.

constructed OPV devices.[24–27] An example of one reported device is shown in Figure 6.4.[23] The details of the device is a pentacene/C_{60} donor–acceptor junction. After the initial optical excitation, the singlet state splits into two triplets rapidly for the case of pentacene.[24] Reports have suggested that this process happens in less than 1 ps.[28] One of the possible combined triplet states can be a singlet. Very large efficiencies of triplets have been reported for pentacene.[24–28] For example, using a fullerene acceptor, a poly(3-hexylthiophene) exciton confinement layer, and a conventional optical trapping scheme, it was demonstrated that a peak external quantum efficiency of 109% at wavelength $\lambda = 670$ nm for a 15 nm thick pentacene film.[23] It was also reported that the corresponding internal quantum efficiency was 160% for this system. This measurement caught many by surprise. Analysis of the magnetic field effect on photocurrent suggests that the triplet yield approaches 200% for pentacene films thicker than 5 nm.[23] Although triplet excitons are dark states, energy may be extracted from them if they are dissociated into charge. This is possible at a junction between pentacene and the fullerene (C_{60}) when the pentacene is oriented approximately perpendicular to the interface. The extraction part of this process has been investigated by several groups around the world.[20]

In considering the materials needed for the process of singlet exciton fission, it is found that the singlet exciton fission (SEF) process may be carried out by either

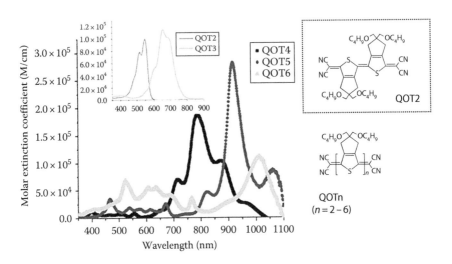

FIGURE 6.5 **(See color insert.)** Organic QOT material and spectra for intramolecular singlet exciton fission. (From Varnavski, O. et al., *J. Phys. Chem. Lett.*, 6(8), 1375, 2015; Chien, A.D. et al., *J. Phys. Chem. C*, 119(51), 28258, 2015.)

an intermolecular and intramolecular mechanism.[1,19,20,29,30] There have been experimental and theoretical investigations of both of these mechanisms. For example, intramolecular SEF has been found to occur in particular carotenoids bound to photosynthetic antenna light-harvesting proteins.[31] It has been also been demonstrated in conjugated polymers.[32] This is accomplished by delocalizing the resulting triplets in well-separated polymer fragments within the maximal conjugation length. These reports inspired scientists to investigate dimers formed by two active molecules covalently connected by electronically inactive units. These systems showed very low SEF quantum yields and slow, noncompetitive triplet formation rates. Another way to approach this is to use strongly coupled carotenoids and polyene-type macromolecular systems, which have demonstrated very high triplet formation rates above 10^{12}/s and the intramolecular SEF yield in certain constructions approaching 30%.[31,32] These systems are strongly correlated with an available multi-exciton state in the vicinity of the first strong one-photon-allowed singlet state (Figure 6.5).

In addition to polymers and polyenes for the discovery of new organic molecules possessing efficient intramolecular SEF processes, the use of quinoidal synthetic structures have also been considered. For example, tetracyanomethylene quinoidal oligothiophenes (QOTn, Figure 6.5) have been reported, giving rise to the possibility of demonstrating singlet fission in one small organic molecule.[1] Long quinoidal oligothiophenes have been shown to possess a biradicaloid character in their ground state. Biradicaloids are known to have low-lying triplet energy levels and therefore are promising structures to meet the energetic requirement $E(S_1) \geq 2E(T_1)$.[1,20,33,34] The quinoidal structure of the thiophene rings results in a mostly planar ground state molecular configuration that undergoes distortions and twisting to stabilize their excited state. The quinoidal features, together with impressive photostability of quinoidal thiophenes, make these structures very promising candidates for a new generation of

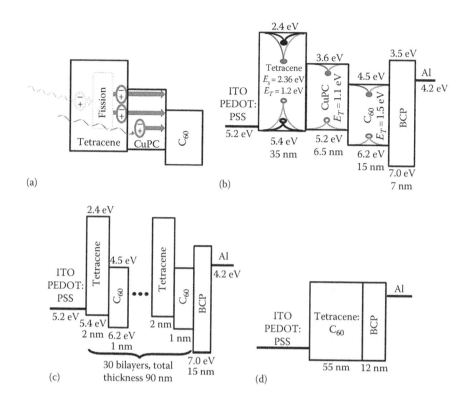

FIGURE 6.6 (**See color insert.**) Device structures. (a) Schematic of a photovoltaic cell exhibiting singlet fission. Tetracene and CuPC are donor materials, and C_{60} is the acceptor. (b) Complete structure of the photovoltaic cell showing singlet and triplet exciton energies and lowest unoccupied and highest occupied molecular orbital energies. Singlets and triplets from tetracene diffuse through CuPC to the CuPC–C_{60} interface. BCP acts as an exciton and hole blocker. (c) Multijunction photodetector structure. (d) Bulk heterojunction solar cell with tetracene–C_{60} blend. (b: From Schaller, R.D. et al., *Nat. Phys.*, 1(3), 189, 2005; d: Chien, S.T. et al., *Laser Focus World*, 10(24), 2013.)

organic molecules for photovoltaics. It was shown that this smaller system gave rise to a triplet efficiency of ~180%.[1,33] This SEF quantum yield is among the largest found for organic molecules featuring SEF and certainly for the intramolecular SEF process. The success of SEF in organic chromophores has given more attention to the role of chromophore coupling in either the intra- or intermolecular process. However, this coupling still requires more theoretical understanding. It is hoped that the processes of extraction and coupling may be given more attention in the future to move these materials closer toward device commercialization (Figure 6.6).

PEROVSKITES AND A NEW HOPE

As mentioned earlier, the field of organic solar cells has experienced intensified efforts in the last 10 years. This is a result of a number of factors having to do with increased pressure to produce efficient solar materials, cheaper device architectures, and the possible decrease in the future cost of silicon-based solar cells. Another factor that has given renewed enthusiasm toward organic photovoltaic (OPV) is the emergence of perovskite-based solar cells.[36-40] While these metal-containing structures are significantly different from its all organic counterpart, the use of organometallic structures still qualifies them as organic hybrid (perhaps semi-organic) based devices. A perovskite is any structure that maintains the same fundamental structure as calcium titinate (ABX_3).[38] These structures have been employed in many applications, and the details of their properties have been well characterized. The reason for the alarming rate of reports and excitement in this area stems from the fact that there have been results of solar efficiencies as high as 20% for some of the prepared perovskite-based solar cells.[41-43] The very high efficiency also comes at a very low cost, mainly due to the relatively simple structure of the organic ligand metal halide structure. New approaches to the fabrication with these structures in TiO_2 and with Al_2O_3 have found that the open-circuit voltage can be impressively large, while the binding energy and thermodynamics are minimized.[44] This approach has provoked questions regarding the long-term goals of the layered organic devices mentioned in previous chapters. In combination with the TiO_2, the perovskite has an amorphous device structure, which absorbs over a very large portion of the visible and near-IR spectrum. It would appear that this material may offer something great in terms of its future application in this field. However, there are a few issues with these systems still left to resolved, mainly the stability of the material when exposed to air.[45] But both scientists and engineers believe this issue can soon be resolved. In this section, we will examine the basic structural and physical model of the perovskite-based solar cells and their properties. We will also look at the important strategies to improve on the long-term stability of the perovskite devices.

PEROVSKITE SYNTHESIS AND STRUCTURE

While we have given many important details of the preparation and properties of all organic solar materials, it is now important in this chapter to relate some of the basic structural properties of perovskites. As mentioned earlier, the structure of these interesting materials follow the repeating formula of ABX_3, where X is either oxygen or a halogen, and A and B are cations that have 12 and 6 coordination with the X halogen. The structural motif is characterized by a tolerance factor, which is a measure of the stability of the final structure in the presence of distortions.[46] In practice, the formation of perovskites used for solar cells have included lead (Pb) and tin (Sn) as the "A" cation and a methylammonium ligand (CH_3NH_3) as the "B" cation. The halide or "X" is typically iodine (I) or bromine (Br).[47] For the typical case of iodine-containing perovskites, one initially starts with a methylamine (CH_3NH_2) in EtOH and adds to this an equimolar amount of hydroidic acid at 0°C for several hours in order to make the methylammonium iodide.[48] The crystallization of CH_3NH_3I can be accomplished through evaporation procedures by adding a small amount of heat over time.[49] In order to obtain the precursor, CH_3NH_3I is mixed with lead iodide (PbI_2) dissolved in solvent and stirred for several hours with heat. The precursor $CH_3NH_3PbI_3$ has been demonstrated to provide a good yield when the preparation is carried out in this manner.[48]

Similar procedures can be carried out for the formation of the Br precursor as well (Figure 6.7).[50] It should be noted that this relatively straight forward approach to synthesis of the pervoskites makes for an excellent educational excercise.

In most publications, the information regarding the synthesis of the perovskite precursor and later material would generally conclude with the information given earlier. But, after the intense interest in these materials over the last 5 years, some do now question what specifically is the chemistry of this process. In particular, what is the reaction mechanism of the formation of the molecular structure and perhaps later the perovskite structure. Very recent reports have set up the chemical equation of the synthesis and elucidate the reactions from both thermodynamic and kinetic perspectives.[51–53] Studies have shown that gaseous products thermodynamically favor the reaction, while the activation energy and "collision" probability synergistically determine the reaction rate.[51] These findings allow researchers to optimize the crystal size for high-quality perovskite films, leading to a record fill factor among similar device structures in the literature.

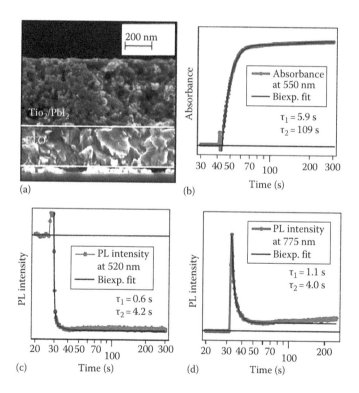

FIGURE 6.7 Microcrystalline formation during the synthesis of PbI_2 containing perovskites. (a) Cross-sectional SEM of a mesoporous TiO_2 film infiltrated with PbI_2. FTO, fluorine-doped tin oxide. (b) Change in absorbance at 550 nm of such a film monitored during the transformation. (c) Change in photoluminescence (PL) intensity at 520 nm monitored during the transformation. Excitation at 460 nm. (d) Change in photoluminescence intensity at 775 nm monitored during the transformation. Excitation at 660 nm. *(Continued)*

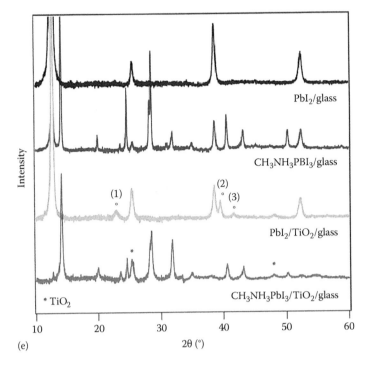

FIGURE 6.7 (*Continued*) Microcrystalline formation during the synthesis of PbI_2 containing perovskites. (e) X-ray diffraction spectra of PbI_2 on glass and porous TiO_2/glass before and after the transformation. The dipping time was 60 s in both cases. The plot shows the X-ray intensity as a function of 2θ (twice the diffraction angle.) (From Burschka, J. et al., *Nature*, 499(7458), 316, 2013.)

$$PbCl_2 + 3CH_3NH_3I \rightarrow CH_3NH_3PbI_{3-x}Cl_x + CH_3NH_3Cl\left(\text{or unknown species}\right)\uparrow$$
$$PbI_2 + CH_3NH_3I \rightarrow CH_3NH_3PbI_3$$

Recent reports have also established the chemical equations for the synthesis of the organic lead trihalide form of perovskite by the two different approaches. Both of the two types of reactions involve the generation of gas species. The process is said to be thermodynamically favorable at elevated temperatures.[53] The kinetics of the reactions are very rapid.[54] Research has found that for the $PbI_2 + CH_3NH_3I$ reaction system, there are rapid collisions. These collisions are critical to the formation and thermodynamics of the products. It turns out that these processes are also important for the film-forming processes as in certain perovskite systems.[54]

MAKING PEROVSKITE FILMS FOR SOLAR APPLICATIONS

It has been shown that perovskites can be incorporated very easily into a standard photovoltaic architecture. While the best perovskite structures have been vacuum-deposited to give better, more uniform film qualities, this process requires the co-evaporation of the organic (methylammonium) component at the same time as the inorganic (lead halide) components.[55] The accurate coevaporation of these materials to form perovskite, therefore, requires expensive evaporation chambers. Some researchers have suggested that this may also cause the practical issues of calibration and cross-contamination between organic and nonorganic sources, which would be difficult to clean.[56] The traditional method for making light-absorbing perovskite films is by blasting it with hot temperatures ranging from 100°C to 150°C. However, the development of low temperature solution deposition routes offer a much simpler method to incorporate perovskites and can even be used with existing materials sets. Although the perovskite solar cells originally came out of dye-sensitized solar cell research, the fact that they no longer require an oxide scaffold means that the field is converging, it seems. Many device architectures now look very similar to thin film photovoltaics except with the active layers substituted with the perovskite. The key to enabling this is that the perovskite precursor materials use relatively polar solvents for deposition; therefore an orthogonal solvent system for the different layers can be fairly easily developed (Figure 6.8).[57]

This structure that results from the synthetic and film preparation procedure discussed previously is a perovskite solar cell based on standard glass/ITO substrates with back metal contact. In order to make a working device from a perovskite film, two charge-selective interface layers for the electrons and holes respectively are needed. Many of the organic photovoltaic interface layers discussed in previous chapters can be used in this process. One of the reasons so many new reports have been published for this material and procedure is, in part, due to the relatively straightforward manner of the synthesis of the perovskite films. For example, it has been reported that such systems as PEDOT:PSS polymers can work as hole interface layers, while C_{60}, ZnO, and TiO_2 make effective electron interfaces.[58]

A great deal of effort has been devoted to optimizing the energy levels and interactions of different materials at the interfaces of perovskite solar devices. Film quality and thickness is of great importance here. It has been suggested that the most efficient perovskite devices are significantly thicker in the active layer than in many organic PV devices. This puts great consideration on the surface roughness or formation of defects in the film. This was initially considered a major problem until higher efficiencies were obtained utilizing spin coating. Now, it is possible to obtain fairly high efficiencies with solution-processed films of perovskites. In connection with this and as stated earlier, a major obstacle in dealing with perovskite films is the issue of stability. In many cases, this relates to the sensitivity to moisture in the atmosphere.[59–65] While the techniques discussed earlier for making films with perovskites do work, they almost all require moisture-free environments to avoid potential humidity-related film degradation. This ultimately requires additional precautions to dry anneal the films, which can lead to additional fabrication costs. New approaches are taking this challenge on by looking at the effects of moisture on perovskite film formation during slow open-air annealing at different temperatures at variable humidity.[62–65] One report compared their air-annealed perovskite films to films grown in traditional dry nitrogen-filled glove boxes.[66] Using atomic force microscopy, it was observed that the films grown slowly in ambient air had larger

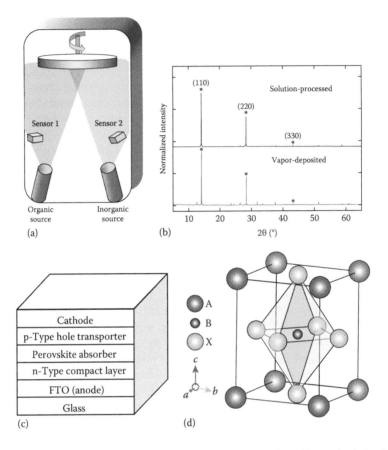

FIGURE 6.8 **(See color insert.)** Perovskite solar cell construction with standard glass/metal substrates. (a) Dual-source thermal evaporation system for depositing the perovskite absorbers; the organic source was methylammonium iodide and the inorganic source PbCl₂. (b) X-ray diffraction spectra of a solution-processed perovskite film (blue) and vapour-deposited perovskite film (red). The baseline is offset for ease of comparison and the intensity has been normalized. (c) Generic structure of a planar heterojunction p–i–n perovskite solar cell. (d) Crystal structure of the perovskite absorber adopting the perovskite ABX3 form, where A is methylammonium, B is Pb and X is I or Cl. (From Liu, M.Z. et al., *Nature*, 501(7467), 395, 2013.)

crystal grains than films grown in a glove box. Larger crystal grains create fewer interruptions for electrons passing through the film, making them more efficient for energy transfer in solar cells. The researchers achieved a maximum ~13% efficiency with their air-annealed perovskite films.[66] Further approaches continue to report on improvements in the thin film fabrication process for perovskites.[67–70] For example, some researchers have reported to develop methods to make perovskite crystal thin films without having to apply heat. Here, a solvent–solvent extraction approach was utilized.[68] This method involves dissolving the perovskite precursors in a solvent and coating onto a substrate. This is followed by the substrate bathed in another solvent, giving rise to apparently a very smooth film of perovskite crystals. Interestingly, there is no heating involved. This process is also very rapid. The researchers suggest that this process can lead to thinner perovskite thin films with high uniformity and could provide a process for mass production.[68]

MECHANISM OF CHARGE TRANSPORT IN PEROVSKITE SOLAR CELLS AND THEIR STABILITY

It is now well established that solar cells composed of organic–inorganic perovskites offer efficiencies approaching that of silicon. But they have been plagued with some important deficiencies limiting their commercial viability. It is this failure that has caught the attention of many scientists and engineers. Researchers at Los Alamos fabricated planar solar cells from perovskite materials with large crystalline grains that had efficiencies approaching 18%.[47] These and other devices are among the highest reported in the field of perovskite-based light-to-energy conversion devices.[42,71] It was reported that the cells demonstrate little cell-to-cell variability, resulting in devices showing hysteresis-free photovoltaic response, which had been a fundamental challenge for stable operation of perovskite devices.[71] The researchers anticipate that their crystal growth technique will lead the field toward the synthesis of wafer-scale crystalline perovskites necessary for the fabrication of high-efficiency solar cells and be applicable to several other material systems plagued by polydispersity, defects, and grain boundary recombination in solution-processed thin films.[71]

The issue of a perovskite's long-term stability is suggestively related to its sensitivity to humidity. Exposure to a humid atmosphere can also significantly change the film morphology. A number of labs have investigated this effect using scanning electron microscopy as described previously in Chapter 3. In one report, perovskite films were deposited on top of a planar poly(3,4-ethylenedioxythiophene) (PEDOT) layer so that changes in the perovskite structure could be clearly observed.[72] They found that as the films age, definite structural changes do occur. This is seen most clearly for the films that were stored at 90% relative humidity. Before humidity exposure, all of the perovskite films have a somewhat rough surface; however, after being in 90% relative humidity for 14 days, the perovskite undergoes a recrystallization process, becoming smooth and highly ordered (Figure 6.9).[72]

As it has been described earlier, lead halide perovskites have recently been used as light absorbers in hybrid organic–inorganic solid-state solar cells, with efficiencies as high as 15% and open-circuit voltages of 1 V. However, a detailed explanation of the mechanisms of operation within this photovoltaic system is still under investigation. The photoinduced charge transfer processes at the surface of the perovskite have been investigated using time-resolved techniques.[73] In a recent report, transient laser spectroscopy and microwave photoconductivity measurements were applied to TiO_2 and Al_2O_3 mesoporous films impregnated with $CH_3NH_3PbI_3$ perovskite and the organic hole-transporting material *spiro*-OMeTAD.[73] It was found that the primary charge separation occurs at both junctions, with TiO_2 and the hole-transporting material, simultaneously, with ultrafast electron and hole injection taking place from the photoexcited perovskite over similar time scales. Charge recombination is shown to be significantly slower on TiO_2 than on Al_2O_3 films. This was the main result that gave researchers the idea that photovoltaic conversion requires two successive steps:

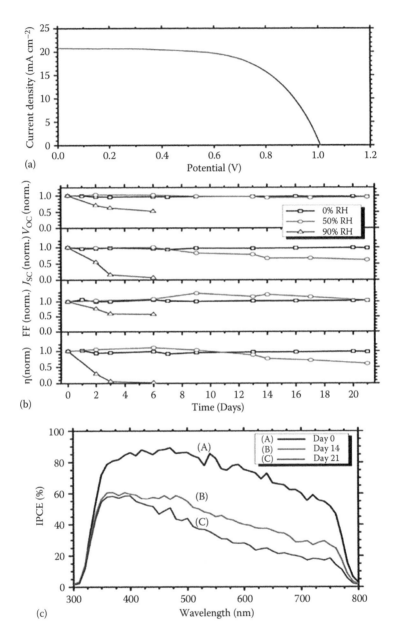

FIGURE 6.9 (See color insert.) (a) Current–voltage characteristics of champion TiO₂/CH₃NH₃PbI₃/spiro-OMeTAD solar cell under 100 mW/cm² AM 1.5 irradiation. The scan was taken with a scan rate of 200 mV/s from forward bias to short-circuit conditions. The active area of the solar cell was masked and the area measured to be 0.113 cm². (b) Stacked plot of the normalized performance parameters with time for solar cells from Table 2 stored in 0%, 50%, and 90% RH. (c) IPCE spectra of perovskite solar cell that was stored in 50% RH taken following exposure to these conditions for (A) 0 days, (B) 14 days, and (C) 21 days.

the accumulation of a photogenerated charge and charge separation of the perovskite at the interface.[73] However, there is still a need for the determination of how and where charge accumulation is attained and how this accumulation can be identified. Both experimental and theoretical studies have probed the mechanism of carrier accumulation in the lead halide perovskite, $CH_3NH_3PbI_3$, thin-absorber solar cells by means of impedance spectroscopy.[74] As it was mentioned in the measurements chapter, this form of spectroscopy can give great details about the electron dynamics in the device and at the interface. In one study, the researchers worked on various cell architectures, using either semiconducting titanium dioxide or insulating aluminum trioxide films. Both porous films were impregnated with lead iodide perovskite ($CH_3NH_3PbI_3$) and an organic "hole-transporting material," which helps extracting positive charges following light absorption. The time-resolved techniques included ultrafast laser spectroscopy and microwave photoconductivity. The results showed two main dynamics. First, that charge separation, the flow of electrical charges after sunlight reaches the perovskite light-absorber, takes place through electron transfer at both junctions with titanium dioxide and the hole-transporting material on a sub-picosecond time scale. Secondly, the researchers found that charge recombination was significantly slower for titanium oxide films than with aluminum systems. As it has been mentioned multiple times before, charge recombination is a detrimental process wasting the converted energy into heat and thus reducing the overall efficiency of the solar cell.[74] Under this context, it is suggested that lead halide perovskites constitute unique semiconductor materials in solar cells, allowing the ultrafast transfer of electrons and positive charges at two junctions simultaneously and the transporting both types of charge carriers quite efficiently. In addition, these findings show a clear advantage of the architecture based on titanium dioxide films and hole-transporting materials.[74]

As one might imagine, perovskite solar cells have rapidly advanced to the forefront of materials which can be processed by solutions and devices with these materials can be quickly made. But, the $CH_3NH_3PbI_3$ semiconductor decomposes rapidly in moist air, limiting their commercial utility. There have been reports of quantitative and systematic investigation of perovskite degradation processes. By carefully controlling the relative humidity of an environmental chamber and using in situ absorption spectroscopy and in situ grazing incidence x-ray diffraction to monitor phase changes in perovskite degradation process, it was found that the formation of a hydrated intermediate containing isolated $PbI_6^{(4-)}$ octahedra as the first step of the degradation mechanism.[72,75] This was significant as it suggested that the identity of the hole transport layer can have a dramatic impact on the stability of the underlying perovskite film, giving way to a route toward perovskite solar cells with long device lifetimes and a resistance to humidity. When exposed to humid air in the dark, $CH_3NH_3PbI_3$ does not revert to PbI_2 with a loss of CH_3NH_3I, as has been postulated for $CH_3NH_3PbI_3$ films at temperature in the light. From these studies and similar detailed characterizations of the process of forming the high efficiency solar device, there is good evidence for the formation of a $CH_3NH_3PbI_3$ hydrate that has similarities to the frequently characterized hydrate $(CH_3NH_3)_4PbI_6 \cdot 2H_2O$.[75] This is a result of analyzing the crystal structure and optical properties of the degraded perovskite material compared to those of PbI_2 as well as $(CH_3NH_3)4PbI_6 \cdot 2H_2O$. The incorporation of

H_2O leads to a loss in absorbance across the visible spectrum. It has been shown that when $CH_3NH_3PbI_3$ is exposed to NH_3 vapor, a structural and optical change occurs wherein the perovskite film turns from dark brown/black to colorless.[65,72,76] In the case of long ammonia exposure (minutes), only a partial recovery of the perovskite structure is possible upon removal of the H_2O vapor is achieved.[75] While significant changes are seen in the UV−visible absorbance, the presence of the hydrate in the films has no noticeable effect on the charge carrier dynamics at short times (<1.5 ns). This is surprising, considering the large carrier diffusion lengths in $CH_3NH_3PbI_3$ and the pervasiveness of trap-state-mediated recombination in semiconductor systems. All of these important findings shed light on the rather complex nature of the stability of the perovskite structure.

It is clear that the crystal structure of the perovskite system is very important. The mechanism of charge mobility and the energetics of the states involved are connected with the structure and can be rather complex. Previous studies have shown by calculation and experiment that defects in the $CH_3NH_3PbI_3$ crystal lattice form only shallow trap states, while trap-mediated recombination mechanisms are essentially absent on the nanosecond time scale.[77] These observations explain the long carrier diffusion lengths of the solution-processed perovskite films and also serve to clarify the effect seen in the present work. It is proposed that H_2O initially reacts with the surface of the perovskite film and only slowly permeates through to the center of large domains.[77] As described earlier, the interface is extremely important, and the connection to humidity is a weak point for the use of perovskites. It is known that the reaction with H_2O at the surface of the perovskite film forms shallow traps in the band structure, so that the portion of the $CH_3NH_3PbI_3$ crystal that is pristine remains largely unaffected.[77] This gives rise to the ultrafast excited state kinetics, and the photo-bleaching peak positions remain unchanged despite the film having undergone near total degradation. An interesting result is that the rate of decrease in photovoltaic performance was faster than the rate of degradation of more molecular-like properties such as the absorption and emission and excited state dynamics.[77] This result is not fully understood. It is believed that many of these results point to a highly interface-driven mechanism for the high efficiency and lack of stability of perovskite films.

LOOKING FORWARD WITH PEROVSKITES AND THE CHALLENGES FACING FUTURE COMMERCIALIZATION

There is now a lot of attention given to perovskite solar cells and their possible commercialization. The resources and enthusiasm seem to be contagious. While it is true that these structures were created relatively recently, perovskite solar cells have shown great promise as an affordable alternative to other solar technologies and their performance achieved new and amazing goals. Reports now suggest that there is a possibility of efficiencies greater than 20% for organometallic perovskite solar devices and upwards of 12% for conjugated polymer devices.[42,78] With the possibility of lower prices for silicon solar devices as a result of this great success, there is now real interest both economically and technologically for hybrid devices.[78]

Indeed, for the structures mentioned in other chapters, it may take more time to reach this milestone since this field started to work on organic structures many years ago. However, recently, the question as to how realistic are these efficiency values from perovskite films has been asked. This question is now being asked internationally, with a cluster of research groups finding that the very nature of efficiency testing, as well as the questionable stability of perovskites themselves, is only serving to exaggerate device performance.[79] And unless this stability problem can be solved, perovskite devices may never become a viable alternative to silicon solar cells. All the exciting efficiencies and any energy claims that are associated with perovskite solar cells should be taken with a grain of salt, according to some scientists. Several international laboratories have been tasked with ensuring that any claims of improved solar cell efficiency are independently verified and held to a robust standard. It is standard practice to test a cell by performing a current–voltage measurement. As mentioned in Chapter 3, the cell is irradiated with 1000 W/m^2 of light at a temperature of 25°C, mimicking the conditions of the midday sun, and its power output is measured. Researchers who test these devices also collect and organize data, which chart the evolving performance of each of the different types of solar cells. From an initial glimpse, this chart appears to affirm perovskites' place as the fastest improving solar technology since silicon.[81] For example, a multicrystalline silicon solar cell, has crawled up the efficiency ladder from about 14% to 20% over two decades. Perovskites, however, have achieved this same increase in just 2 years![81] There may be a simple reason for this distinction, as some believe the instability and rapid degradation of the perovskite films may unintentionally alter the real efficiency measurement provided in most tables.[80] For example, if one would measure in the same manner the perovskite films as they would measure a silicon-based film, it would be clear that the perovskite film would degrade much faster in air. Another factor causing some questions is the fact that the majority of the efficiency values ascribed to perovskite cells have not been independently verified. The uncertainties surrounding perovskite efficiencies do raise a larger issue. If there are concerns about the veracity of these values, who is responsible for ensuring they are verified prior to publication?

In order to have a level playing field and to make good comparisons, we must have a standard form of measurement. As mentioned in Chapter 3, this issue comes up very often. In many cases, those labs that have standard equipment might be the best way to go in terms of having PV cells measured for a reporting efficiency. But, as one would expect, some researchers in this area have been slow to move toward this process. Perhaps it is the journals that might offer some assistance in this regard as well. When is it critical and at what accuracy is it necessary to report the efficiency? From an academic point of view, one would believe that this process is required. However, it appears some groups still do not submit test cells for independent analysis.[82] "This is particularly noticeable for dye-sensitized cells where efficiencies over 13% have been claimed in very high-profile journals, while 11.9% is the best independently confirmed result reported to date".[82] Slowly, for landmark results, some journals are catching on and asking for independent verification. It appears, however, that the perovskite research community is attempting to directly address this problem themselves and to collaborate on these issues[83] There is no doubt that there is a great degree of enthusiasm for perovskites. However, this comes with some caution. In particular, the issue of reporting great efficiencies without mentioning the problem of stability could be problematic.

REFERENCES

1. O. Varnavski et al., High yield ultrafast intramolecular singlet exciton fission in a quinoidal bithiophene, *Journal of Physics Chemistry Letters*, 6(8), 1375–1384, 2015.
2. M. Escalante, A. Lenferink, Y. Zhao, N. Tas, J. Huskens, C.N. Hunter, V. Subramaniam, and C. Otto, Long-range energy propagation in nanometer arrays of light harvesting antenna complexes, *Nano Letters*, 10, 1450–1457, 2010.
3. P.E. Shaw, A. Ruseckas, and I.D.W. Samuel, Exciton diffusion measurements in poly(3-hexylthiophene), *Advanced Materials*, 20, 3516–3520, 2008.
4. V.I. Klimov, A.A. Mikhailovsky, S. Xu, A. Malko, J.A. Hollingsworth, C.A. Leatherdale, H.J. Eisler, and M.G. Bawendi, Optical gain and stimulated emission in nanocrystal quantum dots, *Science*, 290(5490), 314, 2000.
5. M. Beard, Multiple exciton generation in semiconductor quantum dots, *Journal of Physics Chemistry Letters*, 2, 1282–1288, 2011.
6. R.D. Schaller and V.I. Klimov, High efficiency carrier multiplication in PbSe nanocrystals: Implications for solar energy conversion, *Physical Review Letters*, 92(18), 186601, 2004.
7. D.V. Talapin, J.S. Lee, M.V. Kovalenko, and E.V. Shevchenko, Prospects of colloidal nanocrystals for electronic and optoelectronic applications, *Chemical Reviews*, 110(1), 389, 2010.
8. I. Robel, V. Subramanian, M. Kuno, and P.V. Kamat, Quantum dot solar cells. Harvesting light energy with CdSe nanocrystals molecularly linked to mesoscopic TiO_2 films, *Journal of the American Chemical Society*, 128(7), 2385, 2006.
9. O.E. Semonin, J.M. Luther, S. Choi, H.Y. Chen, J.B. Gao, A.J. Nozik, and M.C. Beard, Peak external photocurrent quantum efficiency exceeding 100% via MEG in a quantum dot solar cell, *Science*, 334(6062), 1530, 2011.
10. A.J. Nozik, Quantum dot solar cells, *Physica E: Low Dimensional Systems and Nanostructures*, 14(1), 115, 2002.
11. R.D. Schaller, V.M. Agranovich, and V.I. Klimov, High-efficiency carrier multiplication through direct photogeneration of multi-excitons via virtual single-exciton states, *Nature Physics*, 1(3), 189, 2005.
12. N. Geacintov, M. Pope, and F. Vogel, Effect of magnetic field on the fluorescence of tetracene crystals: Exciton fission, *Physical Review Letters*, 22, 593, 1969.
13. R.C. Johnson, R.E. Merrifie, P. Avakian, R.B. Flippen, Effects of magnetic fields on mutual annihilation of triplet excitons in molecular crystals, *Physical Review Letters*, 19(6), 285, 1967.
14. C.C. Gradinaru, J.T.M. Kennis, E. Papagiannakis, I.H.M. van Stokkum, R.J. Cogdell, G.R. Fleming, R.A. Niederman, and R. van Grondelle, An unusual pathway of excitation energy deactivation in carotenoids: Singlet-to-triplet conversion on an ultrafast timescale in a photosynthetic antenna, *Proceedings of the National Academy of Sciences of the United State of America*, 98(5), 2364, 2001.
15. B. Kraabel, D. Hulin, C. Aslangul, C. Lapersonne-Meyer, and M. Schott, Triplet exciton generation, transport and relaxation in isolated polydiacetylene chains: Subpicosecond pump-probe experiments, *Chemical Physics*, 227(1), 83, 1998.
16. G.R. Fleming, D.P. Millar, G.C. Morris, J.M. Morris, and G.W. Robinson, Exciton fission and annihilation in crystalline tetracene, *Australian Journal of Chemistry*, 30(11), 2353, 1977.

17. A.M. Muller, Y.S. Avlasevich, K. Mullen, and C.J. Bardeen, Evidence for exciton fission and fusion in a covalently linked tetracene dimer, *Chemical Physics Letters*, 421(4), 518, 2006.

18. R.E. Merrifield, P. Avakian, and R.P. Groff, Fission of singlet excitons into pairs of triplet excitons in tetracene crystals, *Chemical Physics Letters*, 3(3), 155, 1969.

19. M.T. Trinh, R. Limpens, W.D.A.M. de Boer, J.M. Schins, L.D.A. Siebbeles, and T. Gregorkiewicz, Direct generation of multiple excitons in adjacent silicon nanocrystals revealed by induced absorption, *Nature Photonics*, 6(5), 316, 2012.

20. (a) M.B. Smith and J. Michl, Singlet fission, *Chemical Review*, 110(11), 6891, 2010; (b) M.B. Smith and J. Michl, Recent advances in singlet fission, *Annual Review of Physical Chemistry*, 64, 361, 2013.

21. I. Paci, J.C. Johnson, X.D. Chen, G. Rana, D. Popovic, D.E. David, A.J. Nozik, M.A. Ratner, and J. Michl, Singlet fission for dye-sensitized solar cells: Can a suitable sensitizer be found?, *Journal of the American Chemical Society*, 128(51), 16546, 2006.

22. L. Salem and C. Rowland, Electronic properties of diradicals, *Angewandte Chemie-International Edition*, 11(2), 92, 1972.

23. D.N. Congreve, J.Y. Lee, N.J. Thompson, E. Hontz, S.R. Yost, P.D. Reusswig, M.E. Bahlke, S. Reineke, T. Van Voorhis, and M.A. Baldo, External quantum efficiency above 100% in a singlet-exciton-fission-based organic photovoltaic cell, *Science*, 340(6130), 334, 2013.

24. W.L. Chan, M. Ligges, A. Jailaubekov, L. Kaake, L. Miaja-Avila, and X.Y. Zhu, Observing the multiexciton state in singlet fission and ensuing ultrafast multielectron transfer, *Science*, 334(6062), 1541, 2011.

25. A. Rao, M.W.B. Wilson, J.M. Hodgkiss, S. Albert-Seifried, H. Bassler, and R.H. Friend, Exciton fission and charge generation via triplet excitons in pentacene/C-60 bilayers, *Journal of the American Chemical Society*, 132(36), 12698, 2010.

26. J.J. Burdett and C.J. Bardeen, Quantum beats in crystalline tetracene delayed fluorescence due to triplet pair coherences produced by direct singlet fission, *Journal of the American Chemical Society*, 134(20), 8597, 2012.

27. S.T. Roberts, R.E. McAnally, J.N. Mastron, D.H. Webber, M.T. Whited, R.L. Brutchey, M.E. Thompson, and S.E. Bradforth, Efficient singlet fission discovered in a disordered acene film, *Journal of the American Chemical Society*, 134(14), 6388, 2012.

28. P.M. Zimmerman, F. Bell, D. Casanova, and M. Head-Gordon, Mechanism for singlet fission in pentacene and tetracene: From single exciton to two triplets, *Journal of the American Chemical Society*, 133(49), 19944, 2011.

29. A.M. Muller, Y.S. Avlasevich, W.W. Schoeller, K. Mullen, and C.J. Bardeen, Exciton fission and fusion in bis(tetracene) molecules with different covalent linker structures, *Journal of the American Chemical Society*, 129(46), 14240, 2007.

30. J.M. Guo, H. Ohkita, H. Benten, and S. Ito, Near-IR femtosecond transient absorption spectroscopy of ultrafast polaron and triplet exciton formation in polythiophene films with different regioregularities, *Journal of the American Chemical Society*, 131(46), 16869, 2009.

31. C. Wang and M.J. Tauber, High-yield singlet fission in a zeaxanthin aggregate observed by picosecond resonance Raman spectroscopy, *Journal of the American Chemical Society*, 132(40), 13988, 2010.

32. G. Lanzani, G. Cerullo, M. Zavelani-Rossi, S. De Silvestri, D. Comoretto, G. Musso, and G. Dellepiane, Triplet-exciton generation mechanism in a new soluble (red-phase) polydiacetylene, *Physical Review Letters*, 87(18), 187402, 2001.

33. A.D. Chien, A.R. Molina, N. Abeyasinghe, O.P. Varnavski, T. Goodson, and P.M. Zimmerman, Structure and dynamics of the (1)(TT) state in a quinoidal bithiophene: Characterizing a promising intramolecular singlet fission candidate, *Journal of Physical Chemistry C*, 119(51), 28258, 2015.

34. J. Wen, Z. Havlas, and J. Michl, Captodatively stabilized biradicaloids as chromophores for singlet fission, *Journal of the American Chemical Society*, 137(1), 165, 2015.
35. S.T. Chien, D.N. Mathews, and M. Grätzel (photo by G. Overton), Perovskite solar cells 5X cheaper than comparable thin-film technology, *Laser Focus World*, 10(24), 24–25, 2013.
36. J. Burschka, N. Pellet, S.J. Moon, R. Humphry-Baker, P. Gao, M.K. Nazeeruddin, and M. Gratzel, Sequential deposition as a route to high-performance perovskite-sensitized solar cells, *Nature*, 499(7458), 316, 2013.
37. M.Z. Liu, M.B. Johnston, and H.J. Snaith, Efficient planar heterojunction perovskite solar cells by vapour deposition, *Nature*, 501(7467), 395, 2013.
38. H.S. Kim et al., Lead iodide perovskite sensitized all-solid-state submicron thin film mesoscopic solar cell with efficiency exceeding 9%, *Scientific Reports*, 2, 591, 2012.
39. J.H. Heo et al., Efficient inorganic-organic hybrid heterojunction solar cells containing perovskite compound and polymeric hole conductors, *Nature Photonics*, 7(6), 487, 2013.
40. H.J. Snaith, Perovskites: The emergence of a new era for low-cost, high-efficiency solar cells, *Journal of Physical Chemistry Letters*, 4(21), 3623, 2013.
41. N.G. Park, Organometal perovskite light absorbers toward a 20% efficiency low-cost solid-state mesoscopic solar cell, *Journal of Physical Chemistry Letters*, 4(15), 2423, 2013.
42. X. Li, D. Bi, C. Yi, J.-D. Décoppet, J. Luo, S.M. Zakeeruddin, A. Hagfeldt, and M. Grätzel, A vacuum flash–assisted solution process for high-efficiency large-area perovskite solar cells, *Science*, 353(6294), 58–62, June 2016.
43. S. Kazim, M.K. Nazeeruddin, M. Gratzel, and S. Ahmad, Perovskite as light harvester: A game changer in photovoltaics, *Angewandte Chemie-International Edition*, 53(11), 2812, 2014.
44. A. Kojima, M. Ikegami, K. Teshima, and T. Miyasaka, Highly luminescent lead bromide perovskite nanoparticles synthesized with porous alumina media, *Chemical Letters*, 41(4), 397, 2012.
45. D. Wang, M. Wright, N.K. Elumalai, and A. Uddin, Stability of perovskite solar cells, *Solar Energy Materials and Solar Cells*, 147, 255, 2016.
46. J.W. Lee, D.J. Seol, A.N. Cho, and N.G. Park High-efficiency perovskite solar cells based on the black polymorph of $HC(NH_2)_{(2)}PbI_3$, *Advanced Materials*, 26(29), 4991, 2014.
47. W.Y. Nie et al., High-efficiency solution-processed perovskite solar cells with millimeter-scale grains, *Science*, 347(6221), 522, 2015.
48. Q. Wang, Y.C. Shao, Q.F. Dong, Z.G. Xiao, Y.B. Yuan, and J.S. Huang, Large fill-factor bilayer iodine perovskite solar cells fabricated by a low-temperature solution-process, *Energy and Environmental Science*, 7(7), 2359, 2014.
49. J.H. Im, C.R. Lee, J.W. Lee, S.W. Park, and N.G. Park, 6.5% efficient perovskite quantum-dot-sensitized solar cell, *Nanoscale*, 3(10), 4088, 2011.
50. M.M. Lee et al., Efficient hybrid solar cells based on meso-superstructured organometal halide perovskites, *Science*, 338(6107), 643, 2012.
51. Q. Chen, H.P. Zhou, Z.R. Hong, S. Luo, H.S. Duan, H.H. Wang, Y.S. Liu, G. Li, and Y. Yang, Planar heterojunction perovskite solar cells via vapor-assisted solution process, *Journal of the American Chemical Society*, 136(2), 622, 2014.
52. B.H. Wang, K.Y. Wong, X.D. Xiao, and T. Chen, Elucidating the reaction pathways in the synthesis of organolead trihalide perovskite for high-performance solar cells, *Scientific Reports*, 5, 10557, 2015.
53. Q. Zhou, Z.W. Jin, H. Li, and J.Z. Wang, Enhancing performance and uniformity of $CH_3NH_3PbI_{3-x}Cl_x$ perovskite solar cells by air-heated-oven assisted annealing under various humidities, *Scientific Reports*, 6, 21257, 2016.

54. A.B. Wong, M.L. Lai, S.W. Eaton, Y. Yu, E. Lin, L. Dou, A. Fu, and P.D. Yang, Growth and anion exchange conversion of $CH_3NH_3PbX_3$ nanorod arrays for light-emitting diodes, Nano Letters, 15(8), 5519, 2015.

55. J. Borchert, H. Boht, W. Franzel, R. Csuk, R. Scheer, and P. Pistor, Structural investigation of co-evaporated methyl ammonium lead halide perovskite films during growth and thermal decomposition using different PbX_2 (X = I, Cl) precursors, *Journal of Materials Chemistry A*, 3(39), 19842, 2015.

56. Y.Z. Li, X.M. Xu, C.G. Wang, C.C. Wang, F.Y. Xie, J.L. Yang, and Y.L. Gao, Degradation by exposure of coevaporated $CH_3NH_3PbI_3$ thin films, *Journal of Physical Chemistry C*, 119(42), 23996, 2015.

57. J.W. Jung, S.T. Williams, and A.K.Y. Jen, Low-temperature processed high-performance flexible perovskite solar cells via rationally optimized solvent washing treatments, *RSC Advances*, 4(108), 62971, 2014.

58. L. Meng, Y.B. You, T.F. Guo, and Y. Yang, Recent advances in the inverted planar structure of perovskite solar cells, *Accounts of Chemical Research*, 49(1), 155, 2016.

59. Y.G. Rong, L.F. Liu, A.Y. Mei, X. Li, and H.W. Han, Beyond efficiency: The challenge of stability in mesoscopic perovskite solar cells, *Advanced Energy Materials*, 5(20), 1501066, 2015.

60. T. Leijtens, G.E. Eperon, N.K. Noel, S.N. Habisreutinger, A. Petrozza, and H.J. Snaith, Stability of metal halide perovskite solar cells, *Advanced Energy Materials*, 5(20), 1500963, 2015.

61. K.A. Bush, C.D. Bailie, Y. Chen, A.R. Bowring, W. Wang, W. Ma, T. Leijtens, F. Moghadam, and M.D. McGehee, Thermal and environmental stability of semi-transparent perovskite solar cells for tandems enabled by a solution-processed nanoparticle buffer layer and sputtered ITO electrode, *Advanced Materials*, 28(20), 3939, 2016.

62. F.K. Aldibaja, L. Badia, E. Mas-Marza, R.S. Sanchez, E.M. Barea, and I. Mora-Sero, Effect of different lead precursors on perovskite solar cell performance and stability, *Journal of Materials Chemistry A*, 3(17), 9194, 2015.

63. J. Schoonman, Organic-inorganic lead halide perovskite solar cell materials: A possible stability problem, *Chemical Physics Letters*, 619, 193, 2015.

64. C. Law, L. Miseikis, S. Dimitrov, P. Shakya-Tuladhar, X.E. Li, P.R.F. Barnes, J. Durrant, and B.C. O'Regan, Performance and stability of lead perovskite/TiO_2, polymer/PCBM, and dye sensitized solar cells at light intensities up to 70 suns, *Advanced Materials*, 26(36), 6268, 2014.

65. G.D. Niu, X.D. Guo, and L.D. Wang, Review of recent progress in chemical stability of perovskite solar cells, *Journal of Materials Chemistry A*, 3(17), 8970, 2014.

66. M.H. Lv, X. Dong, X. Fang, B.C. Lin, S. Zhang, X.Q. Xu, J.N. Ding, and N.Y. Yuan, Improved photovoltaic performance in perovskite solar cells based on $CH_3NH_3PbI_3$ films fabricated under controlled relative humidity, *RSC Advances*, 5(114), 93957, 2015.

67. J.H. Heo, D.H. Song, B.R. Patil, and S.H. Im, Recent progress of innovative perovskite hybrid solar cells, *Israel Journal of Chemistry*, 55(9), 966, 2015.

68. J.H. Heo and S.H. Im, Highly reproducible, efficient hysteresis-less $CH_3NH_3PbI_3$-xClx planar hybrid solar cells without requiring heat-treatment, *Nanoscale*, 8(5), 2554, 2016.

69. B.R. Sutherland and E.H. Sargent, Perovskite photonic sources, *Nature Photonics*, 10(5), 295, 2016.

70. J.M. Frost and A. Walsh, What is moving in hybrid halide perovskite solar cells? *Accounts of Chemical Research*, 49(3), 528, 2016.

71. J. Huang, M.Q. Wang, L. Ding, F.M. Igbari, and X. Yao, Efficiency enhancement of $MAPbI(x)Cl_{(3-x)}$ based perovskite solar cell by modifying the TiO_2 interface with Silver Nanowires, *Solar Energy*, 130, 273, 2016.

72. J.A. Christians, P.A.M. Herrera, and P.V. Kamat, Transformation of the excited state and photovoltaic efficiency of $CH_3NH_3PbI_3$ perovskite upon controlled exposure to humidified air, *Journal of the American Chemical Society*, 137(4), 1530, 2015.

73. A. Marchioro, J. Teuscher, D. Friedrich, M. Kunst, R. van de Krol, T. Moehl, M. Gratzel, and J.E. Moser, Unravelling the mechanism of photoinduced charge transfer processes in lead iodide perovskite solar cells, *Nature Photonics*, 8(3), 250, 2014.

74. H.S. Kim, I. Mora-Sero, V. Gonzalez-Pedro, F. Fabregat-Santiago, E.J. Juarez-Perez, N.G. Park, and J. Bisquert, Mechanism of carrier accumulation in perovskite thin-absorber solar cells, *Nature Communication*, 4, 2242, 2013.

75. J.L. Yang, B.D. Siempelkamp, D.Y. Liu, and T.L. Kelly, Investigation of $CH_3NH_3PbI_3$ degradation rates and mechanisms in controlled humidity environments using in situ techniques, *ACS Nano*, 9(2), 1955, 2015.

76. A.M.A. Leguy et al., Reversible hydration of $CH_3NH_3PbI_3$ in films, single crystals, and solar cells, *Chemistry of Materials*, 27(9), 3397, 2015.

77. J.F. Galisteo-Lopez, M. Anaya, M.E. Calvo, and H. Miguez, Environmental effects on the photophysics of organic-inorganic halide perovskites, *Journal of Physical Chemistry Letters*, 6(12), 2200, 2015.

78. C. Meehan, Perovskite solar cells exceed 20% efficiency, could create hybrid up to 44% efficient, *Solar Reviews (Photovoltaic Technology, Solar Cells)*, 6(15), 1–2, 2016.

79. M. Gunther, Meteoric rise of perovskite solar cells under scrutiny over efficiencies, *Chemistry World*, 3(2), 1–2, 2015.

80. M.A. Green, K. Emery, Y. Hishikawa, W. Warta, and E.D. Dunlop, Solar cell efficiency tables (version 47), *Progress in Photovoltaics*, 24(1), 3, 2016.

81. W. Tress, N. Marinova, T. Moehl, S.M. Zakeeruddin, M.K. Nazeeruddin, and M. Gratzel, Understanding the rate-dependent J-V hysteresis, slow time component, and aging in $CH_3NH_3PbI_3$ perovskite solar cells: The role of a compensated electric field, *Energy and Environmental Science*, 8(3), 995, 2015.

82. E.L. Unger, E.T. Hoke, C.D. Bailie, W.H. Nguyen, A.R. Bowring, T. Heumuller, M.G. Christoforo, and M.D. McGehee, Hysteresis and transient behavior in current-voltage measurements of hybrid-perovskite absorber solar cells, *Energy and Environmental Science*, 7(11), 3690, 2014.

83. S. Meloni et al., Ionic polarization-induced current-voltage hysteresis in $CH_3NH_3PbX_3$ perovskite solar cells, *Nature Communications*, 7, 10334, 2016.

7 A Look Forward with Organic Solar Fuels

THE RIGHT TIME FOR ORGANIC SOLAR TECHNOLOGY

There are images and possible clues of a sustainable solar technology all around us. Due to the great inspiration, ingenuity, and economical awareness of the need for an improved sustainable energy source, there is now very serious consideration for a solar-powered energy economy (Figure 7.1).The creation of organic solar cells has swiftly developed over the last 40 years. There is now a great deal of enthusiasm for finding the key parameters to make these devices commercially viable energy components. While there are still some hard unanswered questions, there is good reason to believe that the present limitations may be overcome. The time for the serious utilization of solar energy has arrived! Present solar technology has come a long way from the first century AD sun rooms appearing in Roman architecture. The new design criteria of recent cells are vastly more sophisticated than the designs and construction of the first solar collectors.

The focus of this book has been directed toward the science and engineering of organic solar devices and to provide the reader with an account of the number of issues, which still remain to be solved. The physics and chemistry surrounding these issues are rather detailed, and the reader is highly encouraged to look through the numerous references of each chapter for further details of the subjects presented. Within this limited frame, the reader can already appreciate the rather complex and broad range of problems and approaches that have been established for the purpose of developing organic solar cell devices. There is a sense of immense excitement as well as moments of uncertainty about the future of organic solar devices. By examining the research in the last few decades in this area, one can notice the periods of high encouragement as well as the challenge of frustratingly slow improvements in efficiency. Perhaps this process is typical in many important advances in technology. With great effort, the field of organic photovoltaic devices has come closer to the projected milestone of 15% efficiency and with a 20-year lifetime to provide electricity at a cost close to seven cents per kilowatt-hour. However, the stability of many of these devices is still a matter of concern. Both efficiency and stability are necessary in order to make organic solar materials competitive with conventional sources of electrical energy. However, it is amazing to see the developments in chemistry, materials science, theory, and engineering that have resulted from the search for new organic solar materials. And these scientific developments have

FIGURE 7.1   (**See color insert.**) Solar fuel at sunset. (Courtesy of T. Goodson III.)

been a strong focus of this book. The increased understanding of the properties of the materials used at the molecular level has led to a more rational exploration of structure–function relationships necessary for improved properties and potential applications.

As stated previously, the goal of this book is to provide a general look at the developments of solar fuels with specific attention to organic materials. Chapters have been devoted to the many successes and failures of several approaches for future materials. While it is agreed that the scope of all solar cells is very broad, this book recognized that the materials used in modern solar cells may be divided into two parts. Historically, it was silicon- and inorganic-based materials that first arrived on the scene in the construction of modern photovoltaic devices. Later, the more aspiring materials used in modern solar cell development were made with organic systems. One reason organic systems came into attention was their ability to be synthetically altered in order to provide direct structure–function relationships. The discussions presented in this book are particularly timely, as this area of research has enjoyed an increased degree of interest and investment due to the potential of high-pay-off technological gains.[1–9] And as the last chapters have demonstrated, not only has the field of organic solar cell discovery enjoyed a number of great accomplishments potentially worthy of translating into the market place, but the field has also developed a deeper understanding of the processes and science involved in their mechanisms. It should not be surprising that the search for the ultimate solar cell material has been saturated with the discussion of efficiencies. While the total cell efficiency is clearly important, this book has tried to focus as well on the physical parameters such as V_{oc}, J_{sc}, and FF in the discussion of different organic solar materials and devices. It is by varying these properties that scientists have been able to improve ultimately on the efficiencies of devices and at the same time learn a great deal about the chemistry and physics of the materials.

ARE THE COMPONENTS OF ORGANIC SOLAR
CELLS ALREADY GOOD ENOUGH?

Over the course of our exploration of organic solar materials, we found that the synthesis and device fabrication of organic solar cell materials can be grouped into two categories. The first category is the synthesis of small molecular systems with high purity. The second group of organic materials for solar applications is comprised of conjugated polymers and other macromolecular architectures with wide bandgaps. And while it was not possible to discuss all of the structures that have been reported, a representative survey of these materials was presented. A major advantage of using small molecules is the relative high purity and control of film thickness and crystallinity in the films. The discussion of the need for organic small molecular structures to have a high efficiency in absorption in the visible and near-IR spectral wavelengths led many researchers to look at structures initially related to porphyrins and pthalocyanines. These materials have the appropriate combination of transition dipole moment and spectral distribution to be used in solar devices. And many of the initial measurements with these organic small molecules helped provide a great deal of structure–function properties useful for further developments. Research in this field has learned how to develop the further use of these small molecules with the formation of layered structures on surfaces. For example, it is now possible for small organic porphyrins to be made into tetramers and other small organic aggregates and later self-assembled onto transparent surfaces as shown in Figure 7.2.[10] This is an exciting development as the field moves even more quickly into the discovery of new molecular architectures as well as device architectures. With the discovery of such impressive organic systems, one might ask the question if the present materials are already good enough for device fabrication?

It was found that smaller donor–acceptor molecules as the active materials in polymeric bulk heterojunction (BHJ) solar cells were excellent candidates for high yield devices.[11,12] By optimizing the growth rate of the films, researchers have

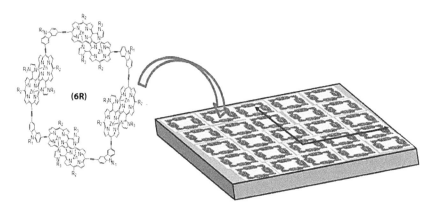

FIGURE 7.2 **(See color insert.)** Square porphyrin structures as a new platform for new surfaces of solar materials. (Courtesy of T. Goodson III.)

worked to improve the performance of the small molecular devices.[12] Donor copoly-mers incorporating various absorbing polymers with benzo[1,2-*b*;4,5-*b'*]dithiophene (BDT) (PTBs) were later found to be particularly suitable for visible, near-IR and IR solar cell operation.[12] Chemistry has been the driving force in the development of structure–function relationships for new solar cell materials. For example, discus-sions of steric effects as well as the planarity of the BDT-type polymers allowed sci-entists to make rational choices of new molecular and macromolecular approaches. It was found that with an extended π-system and great rigidity to the polymer back-bone, better π–π stacking and high hole mobility may be achieved. The PTB and BDT (and many others) have shown great potential. Many researchers find these materials particularly appealing due to the lowering of the bandgap with efficient absorption (around 700 nm) in the region of greatest photon flux of the solar spec-trum. The PTB polymer showed very promising properties for solar cell applications with a maximum in this spectral region and high miscibility with fullerene acceptors. Moving forward, it is expected that the PTB-like polymers will continue to be opti-mized for greater performance.[13] This accomplishment is a result of many scientific and materials investigations of their basic properties. It has already been suggested that some organic light-absorbing materials such as PTB might be at the point of use-ful absorption in the right spectral regions for translational applications.

ORGANIC SOLAR CELL DEVICE CONSTRUCTIONS CONTINUE TO IMPROVE

There is still a great deal of research concerned with the architecture of the device. There continues to be new organic structures for solar applications populating the literature.[14–23] In our exploration of the different organic solar devices we have con-sidered various architectural constructions. The use of concentrators has significantly progressed in the last 10 years as well. It has been suggested that the actual cost of photovoltaic power can be reduced with the use of organic solar concentrators.[24] Our discussion of planar wave guides with a thin-film organic coating on the face and inorganic solar cells attached to the edges led to vast improvements in efficiency. As discussed in previous chapters, these devices may provide very high quantum efficiencies and projected power conversion efficiencies as high as 6.8%. These tech-nologies were reported to enhance the power obtained by photovoltaic cells by a factor of 10. Indeed, there is still room for improvement with this architecture due to the great enhancement obtained. However, this enhancement comes with the cost of increased complexity and manufacturing price. The cost of particular approaches is a real consideration at this point. Most recently, their higher price and lower price-to-performance ratio have limited their use to special roles.[15] We have also observed that the use of tandem techniques can also be used to improve the performance of existing cell designs, although there are strict limits in the choice of materials. In particular, the technique can be applied to thin-film solar cells using amorphous sili-con to produce a cell with about 10% efficiency that is lightweight and flexible. This is an important figure of merit to keep very clear in designing new and potentially commercially viable solar devices. For this purpose, the use of tandem techniques is already in practice by several commercial vendors. There are now techniques that can provide a rapid, scalable, and industrially viable way to manufacture large sheets of flexible organic tandem solar cells. It is hoped that more commercialization of this approach will arrive in the very near future.[16–18] If the roofing solar materials market continues to improve, there could be a substantial market for tandem materials in the United States and in other countries.[19] The reader should now appreciate that there has been a great deal of progress in pushing organic tandem cells toward commercialization. However, we are still left with the difficult question as to what can be done to improve the operation of multilayer systems that will have a large impact on a number of applications for solar power?

Certainly, the first step in characterizing new device constructions is to develop systematic or standard ways of testing devices. While a large number of new solar materials and cell designs are appearing at a very fast pace, the ability to standardize the quality and magnitude of the efficiency is an important parameter. We found in the measurements chapter that the estimate of the device efficiency may not be as easy or as simple as one might have thought. The issue becomes increasingly more complicated for the case of multijunction tandem solar cell devices. We are grateful to such institutions as the Department of Energy, which provides protocols for testing efficiencies, and there are locations around the United States to which one can send particular devices to be evaluated by standardized methods. With the appropriate instruments and techniques, a standardized measurement of the properties of organic PV cells and the associated electronic materials can be carried out.[20] We also learned that the detailed properties of the interfaces can be investigated. This includes such properties as optical, electrical, material, surface, chemical, and structural performance. These characterization tools are particularly important in providing accurately calibrated tools to test efficiencies in novel photovoltaic devices. The performance of PV cells is commonly rated in terms of their efficiency with respect to standard reporting conditions defined by temperature, spectral irradiance, and total irradiance. There is a very specific procedure for this standardized measurement, and it was discussed in the measurements chapter. In a number of chapters, it was discussed that there are now a variety of probe designs available for making contact to PV devices and the fill factor and efficiency can be artificially enhanced by separating the voltage and current contacts or by adding multiple current contacts. The availability of custom test fixtures is useful for the rapid evaluation of PV devices in a production environment. Indeed, the very impressive list of concepts, methods, protocols, and best practices can be very extensive for this field. This is important and suggests that there is still a great deal more to learn in the optimization of a solar cell device construction and performance.

CHALLENGE AT THE INTERFACE STILL REMAINS

As mentioned in several chapters, there are still some important challenges in the use of organic materials for solar applications. One of the important issues that seems to reappear in estimating the limits of organic materials is the process of exciton diffusion.[21] The exciton diffusion length in polymer films is often measured using the diffusion-limited quenching at the interface with either fullerenes or fullerene derivatives. Exciton diffusion lengths are typically in the range of 5–7 nm for many organic materials discussed in the previous chapters. And this intrinsic property of the organic conjugated materials has been the focus of investigations to improve the response in organic materials. New materials and structures are being developed to enhance the exciton diffusion length in organic PV materials.[23–25] In relation to the exciton diffusion of an organic solar device, we appreciate that there is still great interest in understanding the dynamics of important molecular processes in organic solar cell devices. Investigations of processes such as charge transfer and electron–hole pair recombination are still a relatively vigorous area of research for these materials. As much of the processes in the past were modeled by the Shockley–Queisser approach, new ideas have considered the Shockley–Queisser limit further. As it was discussed in previous chapters, there have been some recent improvements that have concluded that the polaron pair recombination rate is a key factor that determines the $J–V$ characteristics in the dark and under illumination. New theory and experiments have started to demonstrate further that these quasi particles are excited and their relaxation or recombination rates provide a key factor in the overall efficiency of organic donor–acceptor-type solar cells (Figure 7.3).[22]

Another major area of further development to be expected with the organic solar cell devices is related to the fact that there is still great enthusiasm for creating new methodologies to improve bulk heterojunctions by the use of different electrodes. As noted previously, organic-based thin-film optoelectronic devices, such as organic solar cells, organic light-emitting diodes, and organic thin-film transistors, have given considerable efforts to perfect this aspect of their devices given the great economic potential that may lead to a new generation of consumer electronic devices. While this point brings great enthusiasm, it is still the case that most optoelectronic devices require at least one electrode with a work function that is sufficiently low to either inject electrons into or collect electrons from the LUMO of a given organic semiconductor. New reports are appearing with new low-WF metals (other than the typical alkaline-earth metals (Ca, Mg) or metals codeposited or coated with alkali elements (Li, Cs) that show great promise.[26] However, the limiting problems of using these electrodes may present new issues, which may require deeper thinking. For example, issues with their use concerns the fact that they are chemically very reactive and easily oxidize in the presence of ambient oxygen and water. Thus, their use in printed electronics presents limitations that can only be overcome by the fabrication of devices in an inert atmosphere and their subsequent encapsulation with barrier-coating technologies, which increases both the cost and complexity of the device architectures. Engineers and scientists are working on these important problems, but there is still a need for considerable shifts from the typical approach of these highly reactive electrodes. There has been some success with the ITO-free polymer electrode solar cell architectures. As the reader may remember, this system is based on the substitution of the ITO electrode by a transparent polymer hole contact, which is supported by metal structures. In order to enable efficient production technologies, the deposition sequence of the electrodes is inverted in comparison to the standard organic solar cell. It is possible that this approach may catch more enthusiasm in the future as a viable approach toward stable electrodes for organic solar cell devices.[27]

In discussing electrodes and exciton diffusion, we come to perhaps the most critical issue of the device physics and efficiency in organic solar cell devices. The characterization of the interface is a very critical issue that still requires more

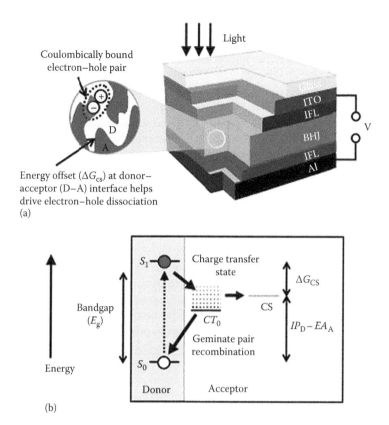

Coulombically bound
electron–hole pair

Energy offset (ΔG_{CS}) at donor–
acceptor (D–A) interface helps
drive electron–hole dissociation

(a)

(b)

FIGURE 7.3 **(See color insert.)** Charge transfer and interfacial effects in organic solar cells. Donor–acceptor energy offsets (ΔG_{CS}) and organic solar cell performance. (a) A bulk-heterojunction (BHJ) organic solar cell showing the Coulombically bound geminate electron–hole pair. Energy offsets at the interface of the donor (yellow) and acceptor (blue) materials drive electron–hole dissociation. ITO = indium tin oxide (a transparent electrode), IFL = interfacial layer, BHJ = bulk-heterojunction absorber layer, and Al = aluminum. (b) Energy landscape at the donor–acceptor interface in an organic solar cell. S_0 is the singlet ground state, S_1 is the singlet excited state, and CT_0 is the lowest energy singlet charge transfer state. The dotted red lines represent the multiple higher energy CT vibronic states. We define the energy offset ΔG_{CS} as $\Delta G_{CS} = E_{g,D} - (IP_D - EA_A)$, where IP_D is the ionization potential of the donor and EA_A is the electron affinity of the acceptor. ΔG_{CS} here represents the lower limit to the thermal energy available for dissociation. Note that triplet states are not shown. The donor and acceptor labels indicate the location of the electron. Note that "geminate pair dissociation" from a CT state to the charge separated state is the focus of the model here. (From Servaites, J.D. et al., *Energy Environ. Sci.*, 5(8), 8343, 2015.)

improvement in organic solar cells. In fact, the design of the electrodes and exciton diffusion is critically dependent on the interactions at the interfaces. Thus, new concepts regarding interface engineering may provide scientists and engineers with a constructive method to facilitate carrier extraction to enhance the organic solar cell efficiency. As the reader should now appreciate, it is known that depending on the chemical structure and properties of the interfacial material, one can strongly affect the polarity of the solar cell device, open-cell voltage, and the contact and thickness properties at the interface. The almost insurmountable list of requirements including (1) the ability to promote ohmic contact formation between electrodes and the active layer, (2) having the appropriate energy levels to improve charge selectivity for corresponding electrodes, (3) having a large bandgap to confine excitons in the active layer, (4) possessing sufficient conductivity to reduce resistive losses, (5) having low absorption in the Vis-NIR wavelengths to minimize optical losses, (6) having chemical and physical stability to prevent undesirable reactions at the active layer/electrode interface, (7) having the ability to be processed from solution and at low temperatures, (8) remaining mechanically robust to support multilayer solution processing, (9) having good film forming properties, and (10) being able to be produced at low cost would suggest we are far from a real solution for a practical interface organic material.[32] But, new measurements and device fabrication schemes have evolved to tackle this list of requirements.[28–31]

Perhaps one of the limitations in previous investigations is the use of the most appropriate measurement to evaluate the physics of the organic solar cell at the interface. As mentioned in the measurements chapter, if one desires a clear physical mechanism of the charge carrier dynamics at the interface, the choice of measurement techniques is very important. New techniques to probe the interface have provided new insights into the structure–function relationships important to efficient charge transfer across the interface. It was mentioned in previous chapters that for some single BHJ photovoltaic cells, efficiencies approaching 10% have been observed. Indeed, if it is possible to tailor the interface of BHJs tandem cells, then the efficiency might also be substantially increased. In considering this possibility, it is now clear that the interface structure and function will be very important in this process. If the issues at the interface could be solved, then one might predict that organic solar cells could have efficiencies comparable to the best cells commercially sold presently.

The reader will also appreciate that in thinking about interface materials, it is not only important to consider the possible increase in efficiency but also to consider the effect of the interface material on the stability and lifetime of the organic solar cell device. The basic science of probing the chemical and physical properties of organic molecules and their interaction with metal electrodes at interfaces is a widely studied area. However, the present theoretical models used to describe the process of electron-hole dynamics in organic semiconductors have been very limited in their true identification of critical structure–property relationships relevant to organic systems. Both scientists and engineers have attempted to develop new models that may one day completely describe the key parameters involved in this relatively sophisticated process in organic materials. For example, we have seen that researchers have suggested the use of electrostatic models, which have the ability to predict the formation of the charge transfer states at the interface with localized carriers. This could be an initial avenue toward a better description. There is now some evidence that has suggested that for the dynamics of exciton dissociation and charge separation in many organic semiconductor donor–acceptor systems, the formation of spatially separated carriers and bound charge transfer pairs strongly depends on the ordering of both the acceptor and donor as well as the order at their interface.[32] The degree of delocalized charge transfer states plays a central role for free carrier generation. As mentioned previously, the excess energy aids in the charge separation process for typical donor–acceptor systems. In the design of organic systems, we learned that it is possible to design donor–acceptor systems with strong delocalized charge transfer character. Indeed, with the proper building blocks, one can construct molecular topologies for enhanced delocalized charge transfer states in oligomers, dendrimers, and polymers. The reader should appreciate how the many pieces of the organic solar cell device puzzle have to fit perfectly in order to solve the problem at the interface. There continues to be new reports investigating the impact of interlayers on injection barriers, built-in fields, surface energy, and surface charge recombination rates in organic solar cells.[33]

WILL NEW APPROACHES CONTINUE TO RISE?

Finally, we saw two new approaches toward organic and organic–metallic solar cells, which have attracted great attention in the field. The exciting area of singlet exciton fission is a spin-allowed process in which two triplets match the energy of one singlet and subsequently lead to the formation of a triplet pair. This can result in a larger number of excitons and later dissociated electrons in the process. There are now a relatively large number of very important reports probing the efficiency of the singlet exciton fission process in organic systems and in constructed OPV devices. Very large efficiencies of triplets have been reported for pentacene. Analysis of the magnetic field effect on photocurrent suggests that the triplet yield approaches 200% for pentacene films thicker than 5 nm. In addition to polymers and polyenes, other organic structures have shown both intermolecular and intramolecular SEF processes with impressive efficiencies. There continues to be further investigations of the role of charge separation and delocalization in the organic films created for the singlet fission process. In order to bring the very promising triplet efficiencies into play for real devices, the dynamics and mechanisms of charge separation are critical. It is certainly hoped that the illustration of singlet exciton fission in this area of research might bring about new and focused enthusiasm for further development and resources for creating organic solar cell devices (Figure 7.4).[34–38]

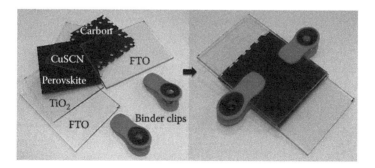

FIGURE 7.4   **(See color insert.)** Perovskite films and devices made for many scientists to appreciate. (From Patwardhan, S. et al., *J. Phys. Chem. Lett.*, 6(2), 251, 2015.)

The second newer area of organic–metallic solar materials is that of perovskite-based solar cells. A perovskite is any structure that maintains the same fundamental structure as calcium titinate (ABX_3). There have been reports of solar efficiencies as high as 20% for some of the prepared perovskite-based solar cells. The very high efficiency also comes at a very low cost, mainly due to the relatively simple structure of the organic ligand metal halide. New approaches to the fabrication with these structures in TiO_2 and with Al_2O_3 have found that the open-circuit voltage can be impressively large while the binding energy and thermodynamics are minimized. However, this approach has provoked questions regarding the long-term goals of the layered organic devices. There are still some issues related to the stability of the material when exposed to the air. Additionally, the lack of standardized testing of these devices has brought about questions about the reproducibility of the impressively high efficiencies reported in scientific journals. Given these limitations, both scientists and engineers still believe these issues can soon be resolved, and indeed these materials are excellent tools in teaching our next generation students about solar devices.[39–43] Time will tell if these approaches will continue to gather such great interest. Certainly, research labs and institutions around the world can provide heavy resources toward solving the current limitations in these new approaches.

FUTURE OF ORGANIC SOLAR CELLS

In considering the future, there is still much to be done in the area of organic solar devices. This is an exciting time. Particularly for students! This is a great time to join this field of research and development and to dive into the chemistry and physics of the details. It brings about a nice combination of science and engineering, theory and experiment, and rational thought. And like most fields of research, there is a touch of necessary good luck as well. What has most researchers excited about the future of organic solar devices is not only the possibility of creating a commercial and cheaper device, but it is the immense basic understanding of the science and engineering of the materials that have evolved already from previous studies. With new and improved scientific techniques and measurements as well as theoretical models, the future of organic solar cell materials is expected to obtain even deeper insights into the physics and chemistry of organic materials. Surely, the future for organic solar cells is bright. From a scientific standpoint, the study of these materials will certainly train many of our next generation scientists and engineers. From a technological and economic standpoint, the use of organic solar cells stands at the crossroads of an important time that will ultimately define the level of effort and resources for serious consideration as a potential sustainable fuel of the future. Only time will tell if solar fuels will be the next generation's ultimate energy resource. Until then, we remain very optimistic.

REFERENCES

1. P. Frankl, E. Menichetti, and M. Raugei, Technical data, costs and life cycle inventories of PV applications, NEEDS project (New Energy Externalities Developments for Sustainability), report prepared for the European Commission, available at www.needs-project.org, 2008.

2. E. Wood, Solar outlook 2015: Still growing, but no longer energy's young kid renewable energy world, January 28, 2015.

3. N. Kaufman, Annual energy outlook projections and the future of solar photovoltaic electricity, Institute of Policy and Integrity, Paper 2014/4, 2014.

4. C. Wadia, A.P. Alivisatos, and D.M. Kammen, Materials availability expands the opportunity for large-scale photovoltaics deployment, *Environmental Science and Technology*, 43(6), 2072, 2009.

5. I. Clover, US retail centers have 62.3 GW of rooftop PV potential, PV magazine (applications and installations Global PV markets), February 2016.

6. A. Momeni and K. Rost, Identification and monitoring of possible disruptive technologies by patent-development paths and topic modeling, *Technological Forecasting and Social Change*, 104, 16, 2016.

7. P. Mir-Artigues and P. del Río, Economics of solar photovoltaic generation, The Economics and Policy of Solar Photovoltaic Generation, Part of the series Green Energy and Technology, pp. 71–159, 2016.

8. P. Denholm, K. Clark, and M. O'Connell, On the path to sunshot. emerging issues and challenges in integrating high levels of solar into the electrical generation and transmission system, Technical Report of National Renewable Energy Lab, NREL/TP--6A20-65800, 1253978, 2016.

9. B. Azzopardi, Future development promise for plastic-based solar electricity, *Progress in Photovoltaics*, 24(2), 261, 2016.

10. O. Varnavski, J.E. Raymond, Z.S. Yoon, T. Yotsutuji, K. Ogawa, Y. Kobuke, and T. Goodson, Compact self-assembled porphyrin macrocycle: Synthesis, cooperative enhancement, and ultrafast response, *Journal of Physics Chemistry C*, 118(49), 28474, 2014.

11. B. Keller, A. McLean, B.G. Kim, K. Chung, J. Kim, and T. Goodson, Ultrafast spectroscopic study of donor-acceptor benzodithiophene light harvesting organic conjugated polymers, *Journal of Physics Chemistry C*, 120, 117, 2016.

12. Y.Q. Zhu, L. Yang, S.L. Zhao, Y. Huang, Z. Xu, Q.Q. Yang, P. Wang, Y. Li, and X.R. Xu, Improved performances of PCDTBT:PC71BM BHJ solar cells through incorporating small molecule donor, *Physics Chemistry Chemistry Physics*, 17(40), 26777, 2015.

13. Q. Wu, D. Zhao, A.M. Schneider, W. Chen, and L. Yu, Covalently bound clusters of alpha-substituted PDI—Rival electron acceptors to fullerene for organic solar cells, *Journal of American Chemical Society*, 138(23), 7248–7251, 2016.

14. N. Chakravarthi, K. Gunasekar, C.S. Kim, D.H. Kim, M. Song, Y.G. Park, J.Y. Lee, Y. Shin, I.N. Kang, and S.H. Jin, Synthesis, characterization, and photovoltaic properties of 4,8-dithienylbenzo[1,2-b:4,5-b′]dithiophene-based donor-acceptor polymers with new polymerization and 2D conjugation extension pathways: A potential donor building block for high performance and stable inverted organic solar cells, *Macromolecules*, 48(8), 2454, 2015.

15. Z. Yu and L.C. Sun, Recent progress on hole-transporting materials for emerging organometal halide perovskite solar cells, *Advanced Materials*, 5(12), 1500213, 2015.

16. X. Li, D. Bi, C. Yi, J.-D. Décoppet, J. Luo, S.M. Zakeeruddin, A. Hagfeldt, and M. Grätzel, A vacuum flash–assisted solution process for high-efficiency large-area perovskite solar cells, *Science*, June 09, 2016, DOI: 10.1126/science.aaf8060.

17. J.J. Intemann, K. Yao, F.Z. Ding, Y.X. Xu, X.K. Xin, X.S. Li, and A.K.Y. Jen, Enhanced performance of organic solar cells with increased end group dipole moment in Indacenodithieno[3,2-b]thiophene-based molecules, *Advanced Functional Materials*, 25(30), 4889, 2015.

18. Y. Zhong, M.T. Trinh, R.S. Chen, G.E. Purdum, P.P. Hlyabich, M. Sezen, S. Oh, H.M. Zhu, B. Fowler, and B.Y. Zhang, Molecular helices as electron acceptors in high-performance bulk heterojunction solar cells, *Nature Communications*, 6, 8242, 2015.

19. L.T. Dou, J.B. You, Z.R. Hong, Z. Xu, G. Li, R.A. Street, and Y. Yang, 25th anniversary article: A decade of organic/polymeric photovoltaic research, *Advanced Materials*, 25(46), 6642, 2013.

20. J.S. Wu, S.W. Cheng, Y.J. Cheng, and C.S. Hsu, Donor-acceptor conjugated polymers based on multifused ladder-type arenes for organic solar cells, *Chemical Society Review*, 44(5), 1113, 2015.

21. M.M. May, H.-J. Lewerenz, D. Lackner, F. Dimroth, and T. Hannappel, Efficient direct solar-to-hydrogen conversion by in situ interface transformation of a tandem structure, *Nature Communication*, 6, 8286, 2015.

22. J.D. Servaites, B.M. Savoie, J.B. Brink, T.J. Marks, and M.A. Ratner, Modeling geminate pair dissociation in organic solar cells: High power conversion efficiencies achieved with moderate optical bandgaps, *Energy and Environmental Science*, 5(8), 8343, 2015.

23. M.A. Green, K. Emery, Y. Hishikawa, W. Warta, and E.D. Dunlop, Solar cell efficiency tables (version 47), *Progress in Photovoltaics*, 24(1), 3, 2016, Published: AUG 2012.

24. A. Polman, M. Knight, E.C. Garnett, B. Ehrler, and W.C. Sinke, Photovoltaic materials: Present efficiencies and future challenges, *Science*, 352(6283), 307, 2016.

25. B.D. Sherman, J.J. Bergkamp, C.L. Brown, A.L. Moore, D. Gust, and T.A. Moore, A tandem dye-sensitized photoelectrochemical cell for light driven hydrogen production, *Energy and Environmental Science*, 9(5), 1812, 2016.

26. A.A. Bakulin, A. Rao, V.G. Pavelyev, P.H.M. van Loosdrecht, M.S. Pshenichnikov, D. Niedzialek, J. Cornil, D. Beljonne, and R.H. Friend, The role of driving energy and delocalized states for charge separation in organic semiconductors, *Science*, 335(6074), 1340, 2012.

27. K.B. Ornso, E.O. Jonsson, K.W. Jacobsen, and K.S. Thygesen, Importance of the reorganization energy barrier in computational design of porphyrin-based solar cells with cobalt-based redox mediators, *Journal of Physical Chemistry C*, 119(23), 12792, 2015.

28. X. Guo, M.J. Zhang, W. Ma, L. Ye, S.Q. Zhang, S.J. Liu, H. Ade, F. Huang, and J.H. Hou, Enhanced photovoltaic performance by modulating surface composition in bulk heterojunction polymer solar cells based on PBDTTT-C-T/PC71BM, *Advanced Materials*, 26(24), 4043, 2014.

29. B. Yang, Y. Yi, C.-R. Zhang, S.G. Aziz, V. Coropceanu, and J.-L. Brédas, Impact of electron delocalization on the nature of the charge-transfer states in model pentacene/C60 interfaces: A density functional theory study, *Journal of Physical Chemistry C*, 118(48), 27648–27656, 2014.

30. Q.Y. Bao, O. Sandberg, D. Dagnelund, S. Sanden, S. Braun, H. Aarnio, X.J. Liu, W.M.M. Chen, R. Osterbacka, and M. Fahlman, Trap-assisted recombination via integer charge transfer states in organic bulk heterojunction photovoltaics, *Advanced Functional Materials*, 24(40), 6309, 2014.

31. M. Oehzelt, N. Koch, and G. Heimel, Organic semiconductor density of states controls the energy level alignment at electrode interfaces, *Nature Communications*, 5, 4174, 2014.

32. W. Tress, N. Marinova, T. Moehl, S.M. Zakeeruddin, M.K. Nazeeruddin, and M. Gratzel, Understanding the rate-dependent J-V hysteresis, slow time component, and aging in $CH_3NH_3PbI_3$ perovskite solar cells: The role of a compensated electric field, *Energy and Environmental Science*, 8(3), 995, 2015.

33. S. Gelinas, A. Rao, A. Kumar, S.L. Smith, A.W. Chin, J. Clark, T.S. van der Poll, G.C. Bazan, and R.H. Friend, Ultrafast long-range charge separation in organic semiconductor photovoltaic diodes, *Science*, 343(6170), 512, 2014.

34. E. Busby, J.L. Xia, Q. Wu, J.Z. Low, R. Song, J.R. Miller, X.Y. Zhu, L.M. Campos, and M.Y. Sfeir, A design strategy for intramolecular singlet fission mediated by charge-transfer states in donor-acceptor organic materials, *Nature Materials*, 14(4), 426, 2015.

35. A.A. Bakulin, S.E. Morgan, T.B. Kehoe, M.W.B. Wilson, A.W. Chin, D. Zigmantas, D. Egorova, and A. Rao, Real-time observation of multiexcitonic states in ultrafast singlet fission using coherent 2D electronic spectroscopy, *Nature Chemistry*, 8(1), 16, 2016.

36. Cooperative singlet and triplet exciton transport in tetracene crystals visualized by ultrafast microscopy.

37. Y. Wan, Z. Guo, T. Zhu, S.X. Yan, J. Johnson, and L.B. Huang, Cooperative singlet and triplet exciton transport in tetracene crystals visualized by ultrafast microscopy, *Nature Chemistry*, 7(10), 785, 2015.

38. S.N. Sanders, E. Kumarasamy, A.B. Pun, M.L. Steigerwald, M.Y. Sfeir, and L.M. Campos, Intramolecular singlet fission in oligoacene heterodimers, *Angewandte Chemie—International Edition*, 55(10), 3373, 2016.

39. T.Y. Zheng, Z.X. Cai, R. Ho-Wu, S.H. Yau, V. Shaparov, T. Goodson, and L.P. Yu, Synthesis of ladder-type thienoacenes and their electronic and optical properties, *Journal of the American Chemical Society*, 138(3), 868, 2016.

40. S. Patwardhan, D.Y.H. Cao, S. Hatch, O.K. Farha, J.T. Hupp, M.G. Kanatzidis, and G.C. Schatz, Introducing perovskite solar cells to undergraduates, *Journal of Physics Chemistry Letters*, 6(2), 251, 2015.

41. J. You, et al., Improved air stability of perovskite solar cells via solution-processed metal oxide transport layers, *Nature Nanotechnology*, 11(1), 75, 2016.

42. S.Q. Huang, Z.C. Li, L. Kong, N.W. Zhu, A.D. Shan, and L. Li, Enhancing the stability of $CH_3NH_3PbBr_3$ quantum dots by embedding in silica spheres derived from tetramethyl orthosilicate in "Waterless" toluene, *Journal of of the American Chemical Society*, 138(18), 5473, 2016.

43. A.B. Djurisic, F.Z. Liu, A.M.C. Ng, Q. Dong, M.K. Wong, A. Ng, and C. Surya, Stability issues of the next generation solar cells, *Physica Status Solidi—Rapid Research Letters*, 10(4), 281, 2016.

Index

A

Absorption, 5–6
  time-resolved, 57
  transient, 10–13, 58–59
  two-photon, 64, 67
Aluminum phthalocyanine chloride (AlPcCl), 103–104
American Society for Testing and Materials (ASTM) Standard, 44, 46
Anisotropy measurements, 10, 60–64
Atomic force microscopy (AFM), 14, 49, 125–126
Azafulleroid system, 32, 34

B

BDT-type polymers, 140
Bilayer organic solar cell, 105–106
Bulk heterojunction (BHJ)
  photocurrent generation, 104–105
  photovoltaic cells, 95, 144
  solar cells, 22–25, 29–30, 32–33, 36, 139–140, 143

C

Charge-coupled device (CCD) detector, 57
Charge separation process, 5–6, 12–13, 31, 97, 103, 127, 129, 145
Charge transport, 5, 13, 36, 57, 102, 104, 127–130
Contact potential difference (CPD), 49
Copper phthalocyanine (CuPc), 26–28, 121
cSFM mode, *see* Electrical conductive scanning force microscopy mode

D

Data acquisition, 57
Degenerated fourwave mixing (DFWM) spectroscopy, 64–65
Delocalization, 62, 65, 97–99, 101, 145
Donor–acceptor molecules, 139–140
Dye-sensitized solar cell, 4, 8–9, 52, 125, 131

E

Electrical conductive scanning force microscopy (cSFM) mode, 9
Energy level tuning, 99–101

Exciton diffusion, 5–7, 10

Exciton diffusion, 5–7, 10
  at interface, 104–106
  length, 80–81, 83, 115, 141–142
  limitations of, 30
  measurement of, 54–57
External quantum efficiency, 47–48, 118–119

F

Femtosecond fluorescence upconversion measurements, 11, 61
Femtosecond transient absorption spectroscopy, 57–60
Fluorescence anisotropy decay, 10, 62–63
Fluorescence upconversion spectroscopy, 10–11, 60–64
Fluorophobic effect, 29, 36
Fourier transform infrared (FTIR), 13

G

Grazing incidence wide-angle x-ray scattering (GIWAXS) studies, 33

H

Highest occupied molecular orbital (HOMO) energy, 26–27, 30–37

I

Impedance spectroscopic measurements, 51–54
Impedance spectroscopy, 10, 129
Indium tin oxide (ITO) electrodes, 76, 96
Integer charge-transfer (ICT) model, 99–101
Interface engineering
  organic solar cells, 95
    delocalized charge transfer states, 97–99
    energy level tuning, 99–101
    exciton diffusion, 104–106
    ICT model, 99–101
    interface layers mechanisms, 95–97
    molecular design, 101–104
    surface control, 107–108
International Electrotechnical Commission (IEC) Standard, 44, 47
Intramolecular singlet exciton fission, 119–120
ITO, *see* Indium tin oxide
ITO-free electrode solar systems, 86, 88
ITO-free polymer electrode, 142

151